HOW IT ALL BEGAN

A Thematic History of Mathematics

Dattatray B. Wagh

University Press of America,® Inc.
Lanham · Boulder · New York · Toronto · Oxford

University Press of America,® Inc.
4501 Forbes Boulevard
Suite 200
Lanham, Maryland 20706
UPA Acquisitions Department (301) 459-3366

PO Box 317
Oxford
OX2 9RU, UK

Printed in the United States of America
British Library Cataloging in Publication Information Available

Library of Congress Cataloging-in-Publication Data

Wagh, Dattatray B.
How it all began : a thematic history of mathematics /
Dattatray B. Wagh.
p. cm
1. Mathematics—History. I. Title.

QA21 .W24 2002
510'.9—dc21 2002032037 CIP

ISBN 0-7618-2410-3 (clothbound : alk. ppr.)
ISBN 0-7618-2411-1 (paperback : alk. ppr.)

™ The paper used in this publication meets the minimum requirements of American National Standard for Information Sciences—Permanence of Paper for Printed Library Materials, ANSI Z39.48—1984

Dedicated to the memory of our youngest son

Sumant Dattatray Wagh

whose untimely death deprived him

of the pleasure of reading this book

Contents

Preface

This book is intended for undergraduate students and teachers of mathematics as well as for all those who want to learn a few interesting stories about the founding fathers of ancient and modern modern mathematics.

The class-room practice today has been taking more and more the "State-the-theorem-and-prove-it" form. Of course, the theorem has to be stated and has to be properly proved. But if a little break in this cut and dry form is effected by adding now and then, a little relevant history of the concepts involved, or a short biographical sketch of the mathematician who discovered the theorem, or an occasional but appropriate reference to the greats in the subject, then the teaching immediately becomes rich and rewarding. This has been my experience and of so many other teachers. This book is intended to provide such enriching material. Learning about the people behind the concepts makes the subject come alive for students. This greatly increases their interest in the subject and of-course, that is the key to effective learning and eventual success.

Mathematics is essentially a western discipline; but not entirely so. Geometry and the Deductive Method came from Greece, numerals and Arithmetic came from India, Algebra from Persia, elementary theory and solution of algebraic equations came from Italy, Analytic Geometry was the work of a French mathematician, Calculus was developed by mathematicians from almost all over Europe including Leibnitz of Germany and Newton of England. The first non-commutative algebra which completely revolutionized the original concept of algebra was created by Hamilton of Ireland.

And all this together, today, constitutes the broad stream of mathematics, a stream tremendously enriched by the smaller streams that came from various countries and joined it.

How did all this happen? How did the smaller streams complementary to the main one grew and got to where the main stream was flowing and how were they accepted by countries and cultures where they did not earlier exist?

How It All Began is an attempt to answer this question. It is a book on the thematic history of mathematics. The emphasis here is on the growth of thinking which built up today's mathematics.

Attempts have been systematically made in the book to highlight problems and challenges which creative mathematicians faced and the ways in which they ultimately resolved them and created mathematics. I firmly believe that exposures to such over-views of the ways of working mathematicians gradually motivate students to read more mathematics and even to create it.

Conceived as a possible first introduction of the students to the subject, the size of this book has been kept small. The choice of topics is restricted mainly to those that touch their syllabi, so that it becomes possible for them to identify themselves quickly with the material in the book.

The material of the book is divided into the following eight chapters:

- *The Great Greek Legacy.* In which is given a brief survey of the work of Euclid. Apollonius and Diophantus and the superb and pace-setting quality of Euclid's Elements.

- *Indian Arithmetic.* Wherein details are given of how Indian numerals and Indian algorithms entered Europe through Arabian and European traders, and how in spite of its superiority over the abacus, it took almost 600 years for Europe to accept it.

- *The Method of Integral Calculus.* Begins with the work of Archimedes in finding expression for areas of a circle and a parabolic segment, his use of the method of double, reductio ad absurdum, followed by Kepler's method for dealing with areas of elliptic sectors, Cavalieri's two Principles which simplify computation of certain areas and Fermat's method of areas, and finally ends with Newton's method of finding areas by the use of a property of the area function which he himself had established.

- *Analytic Geometry of Rene Descartes.* Gives a full survey of La Geometrie of Descartes. Also provides a number of good and serious applications of the Transform-Solve-Invert method which arises out of Descartes' work.

- *The Method of Differential Calculus.* Begins with a reference to the use of maxima and minima in finding normals to conics, the clever use by Fermat of a casual remark of Kepler and finding maximum by use of the method of Differential Calculus, work of Roberval and Descartes, and finally formal formulation by Newton and Leibnitz.

- *The Non-Commutative Algebra of Hamilton.* Gives Hamilton's number-pair definition of complex numbers, his failure in developing an algebra of number-triplets and success in creating Quaternions, the first non- commutative algebra.

- *The Arithmetization of Analysis.* Gives the history of how a metaphysical difficulty arising in the definition of Calculus resulted in straightening out definitions of Limit, Continuity, Convergence, etc. and building up the Real-number system on a sound natural-number basis : includes Cantor's as well as Dedekind's work.

- *The Beginnings of Algebra.* Gives an account of the algebraic work of Al-Khowarizmi, Fibonacci, Cardano, Galois and Emmy Noether; includes excerpts from Einstein's letter to New-York Times giving a value assessment of Emmy Noether's work.

I am aware that a number of other enlivening episodes from mathematics have not been included here. It is so done for two reasons. Since the book was written mainly to provide an insight in mathematics for students and teachers, I have chosen to include in the book only those topics which touch the usually prescribed syllabi. Secondly, I wanted to make the account short and readable. More detailed books on the History of Mathematics are, any way available.

Unlike the usual books on the History of Mathematics, the present book deals only with some select currents in mathematics which have enriched its main stream from time to time. *How It*

All Began concentrates on the beginnings and ways in which parts of mathematics such as Calculus, Algebra and Analytic Geometry grew. Naturally each chapter has a well planned objective and its treatment is accordingly directed.

It must however be conceded that this book is more technical than the other usual books on the history of mathematics. It is so designed to be useful as a text in general mathematics. It would then be necessary to supplement this material with short and informal explanations of such concepts as simple structures in modern algebra, standard equations of conics, Real and Complex numbers, etc. If the student gets his first introduction to mathematics and its central concepts through the history of these concepts, it would be in many ways and for many reasons the most proper introduction. This book will form an excellent supplementary reading for mathematics students at all levels.

This is in short what the present book is designed to achieve – a short interesting introduction to the history of various branches of mathematics, to the lives of the mathematicians who created them, to the problems and the challenges which these mathematicians faced and the ways in which they resolved the same and created new mathematics.

Dattatray B. Wagh

Acknowledgments.

Numerous friends and students reviewed the drafts of this material and helped improve the quality of this book. I would particularly like to acknowledge Professors S. Kapoor of Western Michigan University, M. Murdeshwar of the University of Alberta and S. Pandit of University of North Carolina. I have no words to thank Professor S. Deo of Goa University and Professor V. Lakshmikantham of Florida Tech for their efforts in getting this book published. Finally, I would like to thank our son Meghanad and his family for hosting us at their place in Bethlehem, where much of this work was completed.

Mrs. Hira D. Wagh

CHAPTER ONE

THE GREAT GREEK LEGACY

1. Introduction.

Every civilization requires mathematics to sustain it. This was so even in the olden days of the beginnings of civilization. Some countries developed mathematics just enough to meet their immediate requirements. But some, charmed by the challenging problems of mathematics and excited when they found applications of mathematics to vaster areas of intellectual curiosity such as Astronomy, developed mathematics far beyond their mundane needs. Babylonia, China, Egypt, Greece and India were some of these countries.

It is difficult to decide exactly when and how mathematical ideas first struck the human race. An attempt is made in the next section to visualize how, probably, all this began.

2. Food-gatherers become food-growers.

Like other animals, the humans also were a roaming species. They roamed about gathering what looked like food – fruits on trees, grain on plants and flesh of wild beasts which they killed. They tasted it and tested it to find out if it suited them; selected what suited them and rejected the rest. Gradually as they tasted one food and then another, they came to know what food and which grain suited them best as staple food. They identified the plants which grew these grains, the trees which grew these fruits and the beasts which gave them the flesh which agreed with them. Since these plants provided them with their greatest necessity, they carefully noted other features of these plants; where and in what soil and in what surroundings they grow; under what stars in the sky they start growing; and under what stars they start bearing grain. In this process they observed and noted that after a few days, these plants stop growing, become ripe, bear grains and then wither away. They observed that this cycle was perpetual. They were happy to discover it. It meant that these plants were capable of providing them food year after year with almost nothing for them to do except harvesting the grain when it was ready.

They must have noticed that these grain-plants grew where there was water near-about; on banks of lakes or in river basins. The humans also required water, for drinking and washing. These rivers and these lakes could provide water to them, their banks could provide them grain. Where was then the need now, they must have thought, to roam about in search of food! Why not settle down on the banks of these very rivers which fulfilled two of their major necessities of life, for which, till then, they were roaming about all the year round.

But they were used to roaming. It had become their second nature. What would they be doing if they were not roaming? But that was a smaller question. Considerations of security of food and drink were certainly more important. They could not ignore them. Naturally these considerations finally prevailed. They decided to settle down. With it their life-style changed. Earlier they were *gathering* food; now they started *growing* it.

Food-gatherers became food-growers.

It was a major change. For thousands of years they were living their lives in one routine style: rest at a place for a little while; go around the place where they stopped; collect whatever food was available in the region; and equipped with it go ahead to find a similar another place. Stay at this new place for some time and then again go ahead.

This was their routine for a very very long time; for thousands of years, during which phase they made a number of important discoveries in respect of the food that suited them best, the plants which produced this food, the soil conditions in which these plants grew and such other thousand and one small and large things. It was only after such a long experience that they settled down on the river banks. Their old routine changed.

But the problem of their food was not over just on this account. Though they now became food growers, it did not stop their roaming about entirely. Earlier they were roaming about in search of food. Now they had to do it in search of land. A new routine started. Even this second routine continued for thousands of years.

To start with, there was no difficulty in this second phase. Land was plentiful : and comparatively the settlers were few. But then a time came when the population of the settlers grew quite large while new land was scarce to find. And the usual problem faced them : of scant resources and plenty consumers. The only solution in this case was to

divide the land among the rapidly increasing number of land users.

This was a new problem for them. It involved the concept of *size* of land-pieces. They could recognize that one heap of grain was bigger than another or that one beast which they killed was larger than another and would give them more flesh. But to arrive at a judgment regarding sizes of land-pieces was another problem: difficult as well as new. For a time, they used various tentative measures for the purpose. For example, if there were two land pieces and they had to decide which one of them was the larger one, they would walk along the boundaries of the two pieces and conclude that that piece was larger whose boundary took more paces to cover. But they would soon realize that this is not correct. The area of a 10 by 1 rectangle is 10 and its perimeter is 22; while the area 16 of a square of side 4 is larger than the area 10 of the rectangle, but its perimeter 16 is smaller than the perimeter 22 of the rectangle. They also used other tentative criteria all of which later proved to be wrong.

But the problem was urgent for them. Even when they found that all of their tentative solutions were not tenable, they could not give up their search for a correct solution. And though this search did not enable them to arrive at a correct solution of the problem on hand, their continuous efforts brought to their notice a number of facts about areas and about what we now call geometrical figures.

Thus consideration of questions regarding sizes of land-pieces led them to the rudiments of geometry.

Their decision to settle down at one place raised other problems also. They had to *build* houses to stay and granaries to store the grain which they produced. They had to divide the grain amongst the different groups in the settlement. They had to *count* how many members the groups had. Considerations and practical solutions of these problems led them to the rudiments of arithmetic, engineering and other mathematical disciplines.

3. Egyptian mathematics.

This is what must have happened in Egypt in the settlements in the basin of river Nile. An Englishman, A. Henry Rhind, in 1858 secured an Egyptian text which is now in the British Museum. This papyrus, since called the *Rhind Papyrus*, was composed in about 1650 B.C. and contains most of the mathematics developed till then in Egypt. The text contains about 80 problems with solutions of many of them. The problems are mainly of a practical nature. Some are arithmetical, some algebraic and some geometrical.

Some of the methods which they have used in their work are interesting and novel. They demonstrate, in a way, the variety of ways in which human genius from different parts of the world looked at the same problems. It is like different mathematicians giving different proofs of the same theorem.

Let us come back to Egyptian mathematics. Consider the novel way which they employed in the multiplication of two natural numbers. Consider the problem

Multiply 57 by 73.

To do this, they did three simple operations: doubling a number, dividing a number by 2 and adding a set of numbers. If the number, m to be divided by 2 was odd, they would simply divide $m-1$ by 2 and take it as the half of m.

Thus to multiply 57 by 73, they would write 57 and its successive halves in one column and 73 and its successive doubles in another; ignore those doubles which appeared in the rows of halves which were even numbers and add the rest of the doubles. The sum so obtained is the required product. Thus in the multiplication of 57 by 73, the calculation would appear as shown below. The num-

57	73	
28		146
14		292
7	584	
3	1168	
1	2336	
sum	4161	

bers 146 and 292, of the column of the doubles are written aside because they correspond to the *even* numbers 28 and 14 of the column of the halves and as such are not to be added. The sum 4161 of the rest is the product of 57 and 73.

The method is remarkable. And it is more remarkable that this ingenious method should strike the Egyptians 3000 years ago. Today we can easily justify it with the help of the binary system of numbers.

The Egyptian formula

$$(d - \frac{1}{9}d)^2$$

for the area of a circle of diameter d is similarly praiseworthy. It is understandable that it is not correct. But even arriving at such a simple and fairly good approximation for the area of a circle is no mean achievement.

As elsewhere, the Egyptians have only given the formula; but no explanation about how they arrived at it. Experts in interpreting Egyptian mathematics have suggested the following as a probable way in which the Egyptians arrived at the formula.

> Let the circle in the figure above be of diameter d. Let square ABCD of side d circumscribe the circle. Let the

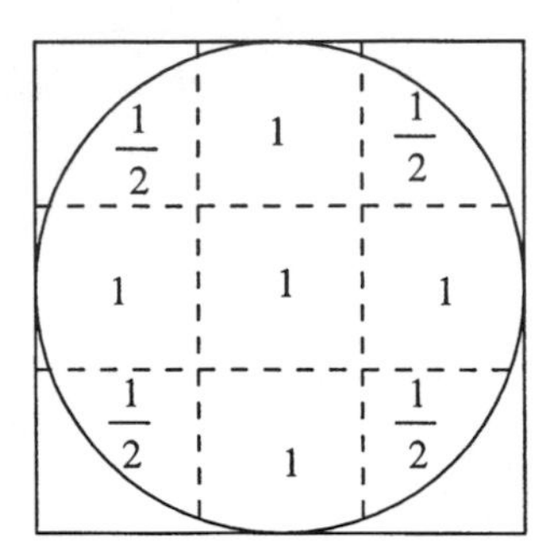

square be divided into 9 smaller squares. The circle consists roughly of 5 full smaller squares and 4 half smaller squares; that is, altogether of 7 full smaller squares. Since the area of each smaller square is $(1/9)d^2$, the area of the circle amounts to

$$7 \cdot \frac{1}{9} d^2$$

which is equal to

$$\frac{63}{81} d^2.$$

This is approximately $(64/81)d^2$. That is $(\frac{8}{9}d)^2$, or as Egyptians gave it

$$\left(d - \frac{d}{9}\right)^2.$$

4. The beginnings of Greek Geometry.

This work of the Egyptians came to the notice of the English archivist Rhind in 1858. It must surely have come to the notice of the Greeks much earlier. They must have been highly impressed by it. The Egyptian interest was practical; they wanted to solve problems which faced them. The Greeks was a race of intellectuals.

They did not look at Egyptian work as work which would be of help to them in the solution of their problems. The interest of the Greeks was purely intellectual. They were extremely excited by the Egyptian work in geometry. When they looked into it they found that not all of it was correct. Formulae for areas of a rectangle and a triangle were correct, but not any of the rest. And the Greeks wondered at both: why some are correct and how Egyptians had arrived at these correct results; and why some other results of the Egyptians were incorrect and where, in these latter cases had the Egyptians gone wrong. Was it because there were errors in their computations or errors in their arguments or errors in some of their assumptions or was it because Egyptian work lacked arguments and reasoning altogether and consisted of only intelligent guesses?

The Greeks decided to study the whole problem *ab initio* and put things straight.

This was the start of Greek Geometry and Greek mathematics in general.

5. The Greek Geometry.

Greek Geometry came down to us as a fully perfected package. There is no point in dissecting it and pointing out that this bit is a contribution of Thales, and that bit is a contribution of Pythagoras and that the other bit is the contribution of Euclid. Greek Geometry is just one piece of exciting material; a superbly organized piece of perfect logic.

The Greeks never cultivated geometry for any material gain. Plato was furious about it. Plutarch remarks upon "Plato's indignation of it and his invections against it as corruption and annihilation of the one good of geometry which was thus shamefully turning its back upon the unembodied objects of pure intelligence." But if they discovered that lack of proper thinking or tolerance of tentative assumptions led the Egyptians or the Babylonians to mistaken notions or incongruent results, the Greeks would pick them up as a challenge to their intelligence and with their habits of disciplined thinking put them straight.

This might have made them pick up with priority, the problem of the areas of geometrical figures. The casual manner in which the Egyptians handled it was enough of a challenge to the Greek mind.

The first decision which the Greeks took in respect of comparing areas of such figures as Fig. A and Fig. B below was:

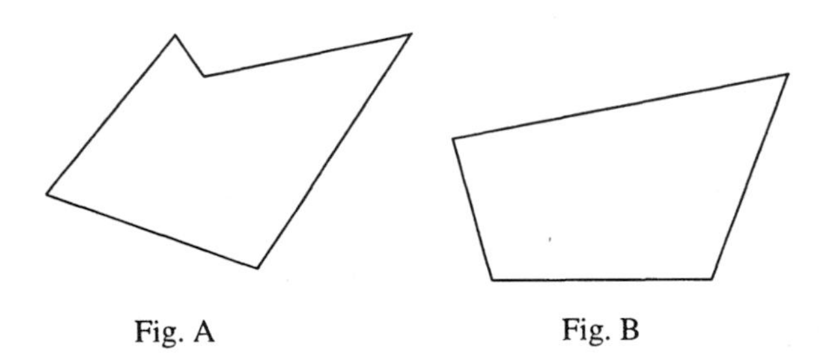

Fig. A Fig. B

> Their area can be compared if there is *one common unit* in which the two areas can be converted.

This led to a second natural question:

What can this common unit be?

In order to arrive at a unit of the desired type, they gave a careful thought to all the regular shapes with which they were familiar, namely, the rectangle, the parallelogram, the triangle, the square and the circle. And they noticed, a little to their surprise, that of

these shapes, the *square* and the *circle* have the special property that their

sizes determine their dimensions

in the sense that there is only *one square* of a given area, the same being also true for a circle. (This is not so for a rectangle; rectangles 6 by 2, and 4 by 3; both have area 12). This naturally makes the *square* and the *circle* good units in terms of which one may measure sizes of geometrical figures. Apart from the fact that the square is easier to handle than a circle, the square has one other point of advantage over the circle. Two squares can be easily added to give a third square as their sum. With circles also this can be done, but with less ease. They thus arrived at the decision that

> *Given geometrical figures should first be transformed into equivalent squares, and then their sizes be compared.*

6. Squaring of rectilinear figures.

But how does one do it? How could one construct a square equal to a given rectangle, or to a parallelogram, or to a triangle or to any other rectilinear figure or to a circle?

This process of constructing a square equal in area to a given figure is called

squaring the figure.

Before anything else, they fixed a unit to measure areas. If AB is a unit length then they fixed that a square of side equal to a

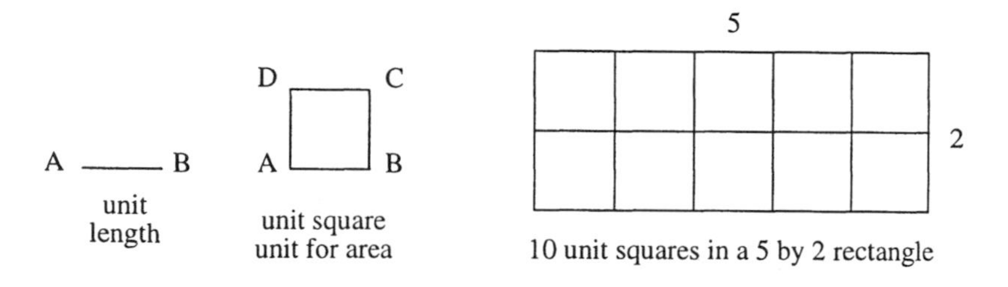

unit length be *a unit square*, a unit to measure area. As is seen in the above figure, a rectangle of sides of 5 and 2 unit lengths gets broken into 10 unit squares. Thus the area of a p by q rectangle is pq unit squares.

Thus they came to two basic decisions:

1. The unit square should be a unit of area, and

2. Areas of geometrical figures be compared after squaring them.

The actual squaring of rectilinear figures is shown below. Squaring a rectangle is the basic construction here. Because Euclid has shown in his *Elements* how a rectangle can be constructed equal in area to a given parallelogram or to a given triangle. Euclid's propositions in this matter are:

Prop. 1. Parallelograms on the same base and between the same parallels are equal in area.

According to this proposition, the area of parallelogram

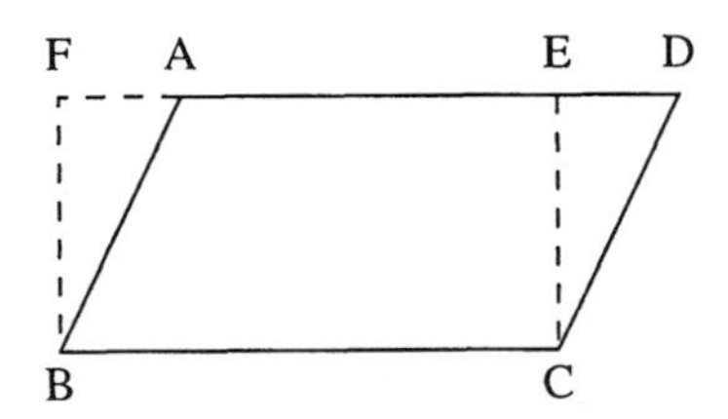

ABCD is equal to the area of rectangle BCEF.

Prop. 2. A triangle is equal in area to a rectangle on half its base and of the same height.

According to this proposition, the area of triangle ABC is equal to the area of rectangle BEFG where E is the midpoint of BC.

Since the above theorems help constructing rectangles equal in area to given parallelograms or given triangles, what matters is a method to square a rectangle. Because of Euclid's propositions given above this should suffice to square parallelograms and triangles.

We can now show how the Greeks squared rectangles.

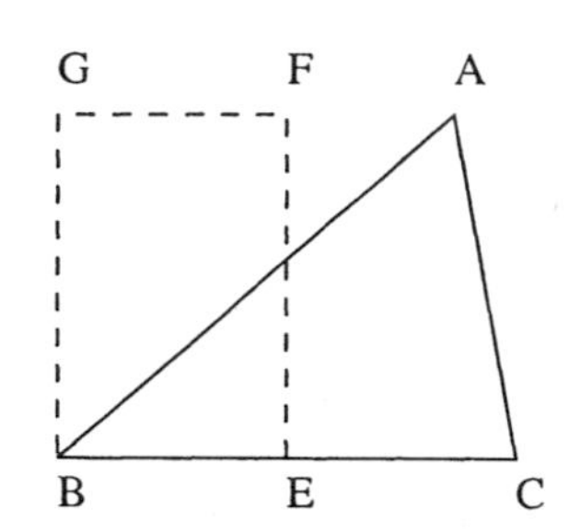

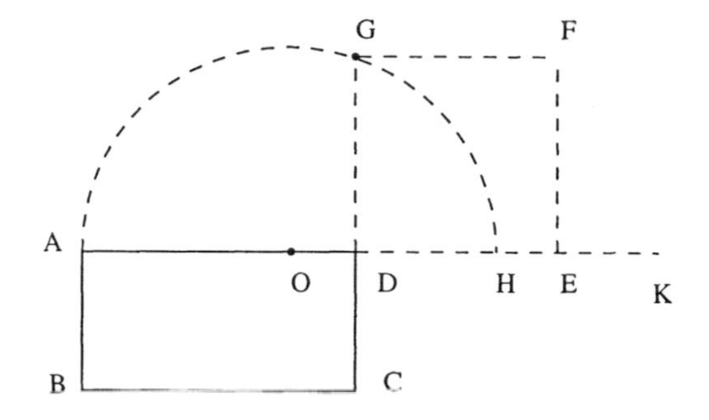

> Suppose, rectangle ABCD is to be squared. To do this, first produce AD to K and on AD produced choose a point H such that DH = DC.
> On AH as diameter, draw a semicircle to meet CD produced in G.
> Complete the square GDEF.
> This is the square equal in area to rectangle ABCD.

Rectangle ABCD has thus been squared.

As a second problem, we shall see how polygons can be squared. Let us start with a simple problem of this kind, that of

squaring a quadrilateral.

To square a quadrilateral, we first *triangulate* it. That is, we break the quadrilateral into triangles by joining some one of its vertices

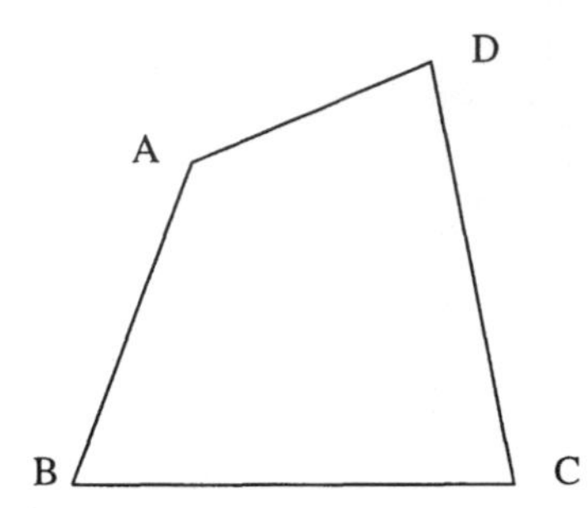

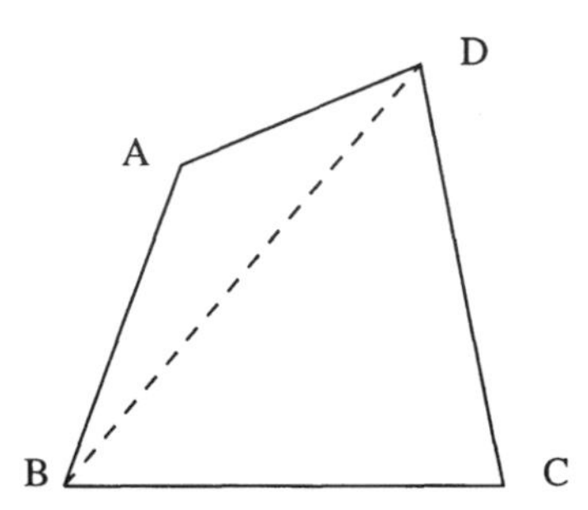

to all the others as shown above. Here we join vertex B to vertex D, which, in this case, is the only vertex not already joined to B. Thus we break the original quadrilateral ABCD into *two triangles*, ABD and DBC. We now square these triangles.

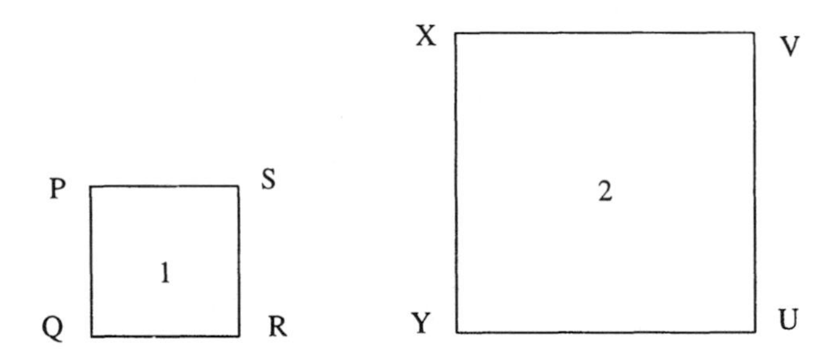

Suppose that when squared, the squares which we get are squares PQRS and XYUV equal in area respectively to triangle ABD and triangle DBC. In order to get one square equal in area to the quadrilateral ABCD, we add these two squares. To do this we make side RS of the first square an extension of side YX of the second square. We now join S to V and construct square SVWT on SV as a side. This construction is shown in the figure on the next page.

Now according to Pythagoras theorem,

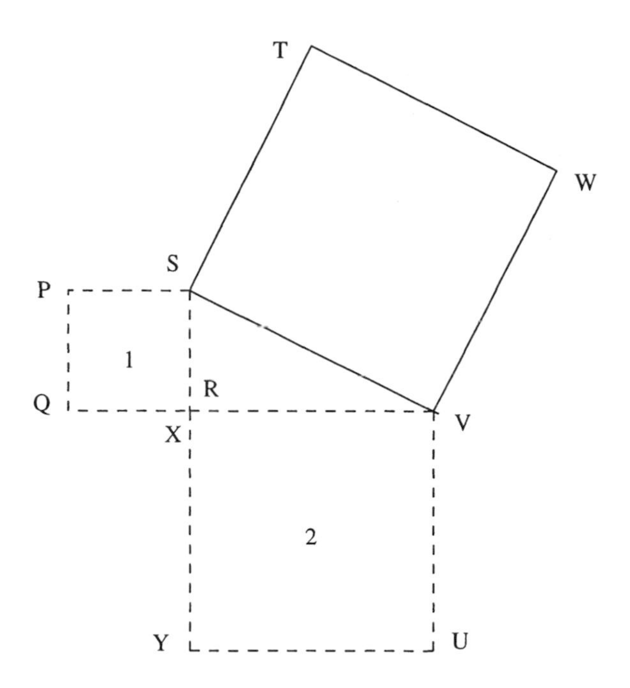

area of square SVWT = area of square 1 + area of square 2
= area of triangle ABD + area of triangle DBC
= area of quadrilateral ABCD.

which means that quadrilateral ABCD has been squared.

Even if the polygon to be squared is a polygon of more sides as shown in the figure on the next page, the procedure to square it is the same as for a quadrilateral. The first step would be triangulation of the polygon, which in the present case, would break the polygon into four triangles. Now each of these triangles could be squared, giving us four squares 1, 2, 3, 4 as shown, equal in area respectively to the four triangles 1, 2, 3, 4 into which the polygon has been broken because of the triangulation.

Repeated application of Pythagoras theorem would now give us one square as the sum of these four squares. This last square is

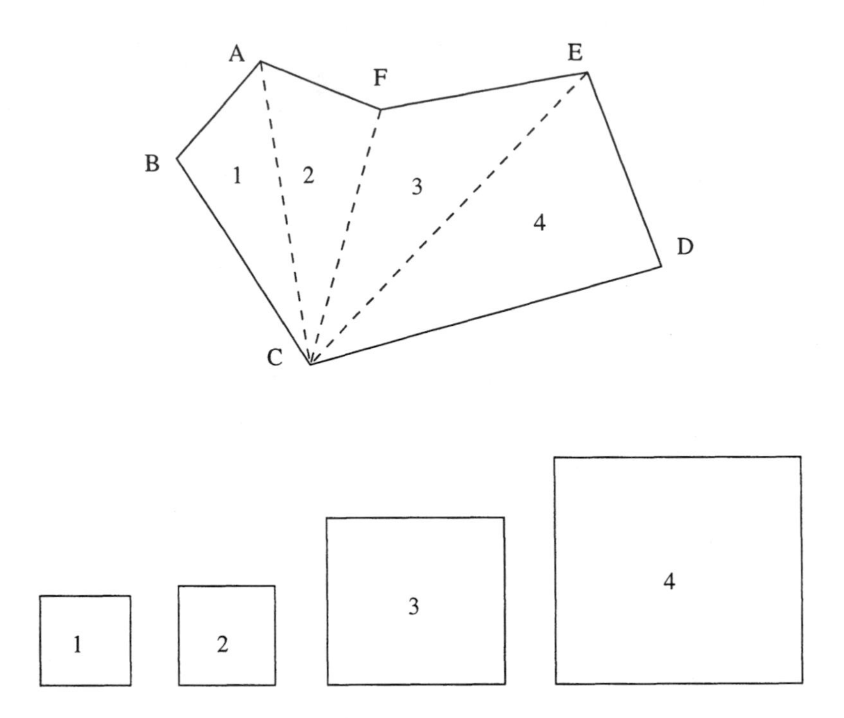

equal in area to the given polygon. The polygon is thus squared.

This method obviously would resolve the question of squaring any rectilinear figure. The figure in the squaring of which this method does not succeed is the circle. The squaring of the circle is discussed in the next chapter.

Incidentally, the method of squaring polygons brings out the basic geometrical use of the Pythagoras theorem. This use of the theorem is unavoidable. Solving squaring problems at the earliest opportunity was necessary; because, area is one of the basic aspects of the study of closed geometrical figures. Therefore, the squaring procedures had to be included in the earlier parts of geometry. Since the Pythagoras theorem is an integral part of this procedure, the theorem had to be established early. This was probably the reason why Euclid included it in his first book where he was forced to give a clever but not pleasant proof of it. If he could have waited till Book VI he could have given the much simpler proof of the theorem based on properties of similar triangles.

7. Number Theory in the *Elements*: Preliminary notions.

The three books VII, VIII and IX of the *Elements* contain a total of 102 number-theoretic propositions. Some of them play a central role in Number-Theory even to this day. For example, the algorithm designed by Euclid to find the greatest common divisor of two natural numbers, contained in propositions VII.1 and VII.2, known now as the Euclidean Algorithm, is in use today not just for the purpose for which it was originally designed but also for various other important purposes. The same can be said about the basic number-theoretic proposition VII.30 that if a prime p divides a product $a \times b$ of two natural numbers a and b then p necessarily divides at least one of a and b. The famous theorem that the number of primes is infinite is a discovery of Euclid and is given in proposition IX.20 in the *Elements*. Its extremely neat and elegant proof, also a discovery of Euclid, has earned for him the highest laurels from many of the great number-theorists of today. The proof has been used effectively as model for proofs of many similar results.

Not all the 102 propositions of these three books are of the same status. Some are just minor; they fill the gaps. And some others are of the nature of lemmas which assist in proving the bigger theorems.

A question is often asked: if the number-theory material included in these three books is so important, then why did Euclid give it in a book predominantly devoted to geometry? Why did he not make a separate book of it?

There are two reasons for this. Firstly, in Euclid's time, there was *no algebra* in Greece. There was neither symbolism sufficiently developed to handle material of this intricate nature nor any established methods or algorithms to deduce one result from another. But besides this, there was a more subtle reason why the material has been given in the language of geometry and proved by methods of geometry. The reason lay in the

> Greek mind which would not accept the truth of any mathematical assertion unless the same was demonstrated geometrically.

In fact, this is not the first place in the *Elements* where algebraic results are demonstrated geometrically. Earlier in the book, Euclid demonstrates the truth of the algebraic result

$$a(a+b) = a^2 + ab$$

from the figure

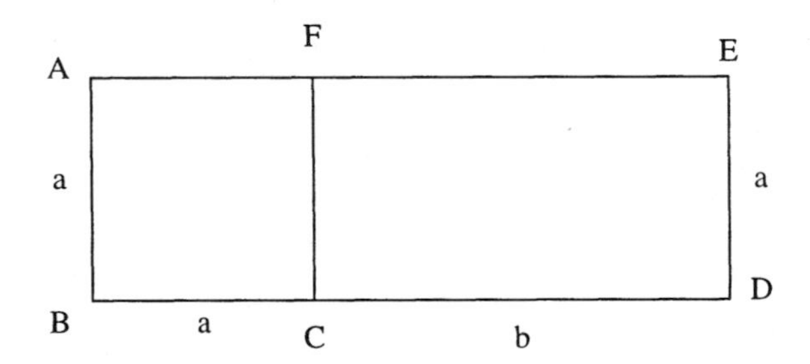

where rectangle ABDE whose sides are a and $(a + b)$ has area $a(a + b)$.

Because of line CF, this rectangle falls into two parts, a square ABCF of area a^2 and a rectangle CDEF of area ab.

Since area of rect. ABDE = area of sq. ABCF + area of rect. CDE

$$a(a+b) = a^2 + ab.$$

The *Elements* gives a similar geometric demonstration of the formula

$$(a+b)^2 = a^2 + b^2 + 2ab$$

by arguing similarly as above with reference to the figure on the next page, where ACEG is a square of side $(a+b)$ and has an area

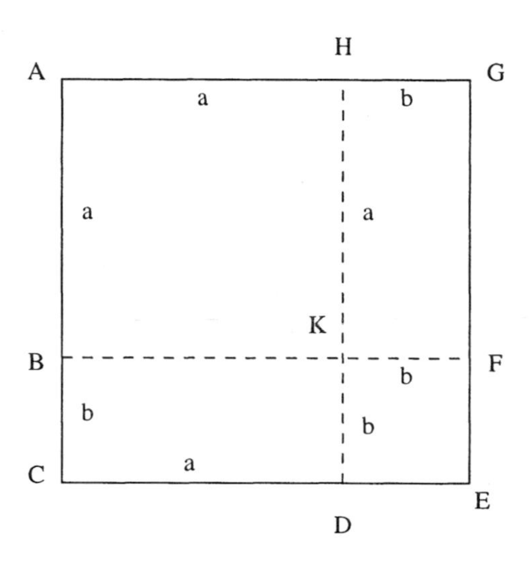

$(a+b)^2$. The lines BF and DH intersecting in K, break square ACEG into the four parts

sq. ABKH, rect. BCDK, sq. DEFK and rect. FGHK

whose areas are a^2, ab, b^2, and ab respectively. Equating the area of the whole and the sum of the areas of its parts, one gets

$$\begin{aligned}(a+b)^2 &= a^2+ab+b^2+ab \\ &= a^2+b^2+2ab\end{aligned}$$

as was desired to be demonstrated.

The geometry which was employed in proving such results was not required to be in any way distinctive. But Number-Theory is an algebra of *natural* numbers. Euclid had to show great ingenuity in devising such geometrical representation for natural numbers that the geometrical results which were then established in respect of them had a suitable interpretation in the algebra of natural numbers. What he did was following:

He decided that

1. Every natural number A be represented by a segment

2. To *add* two natural numbers A and B, the segment XY representing A be produced in its own line to a point Z such that YZ = B, so that XZ represents A + B.

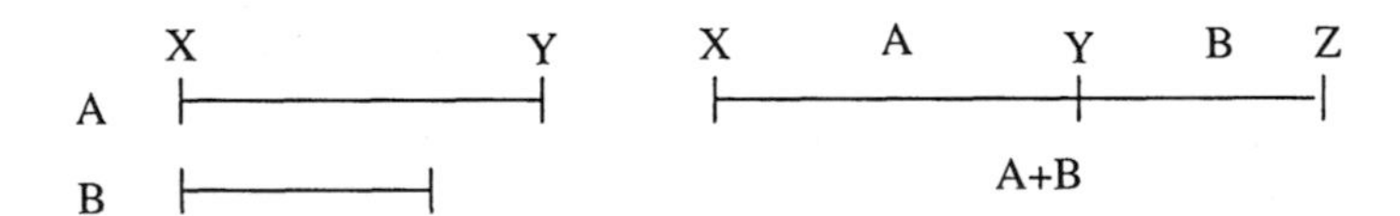

3. To *multiply* natural number A by natural number B, the number A be added to itself B times.

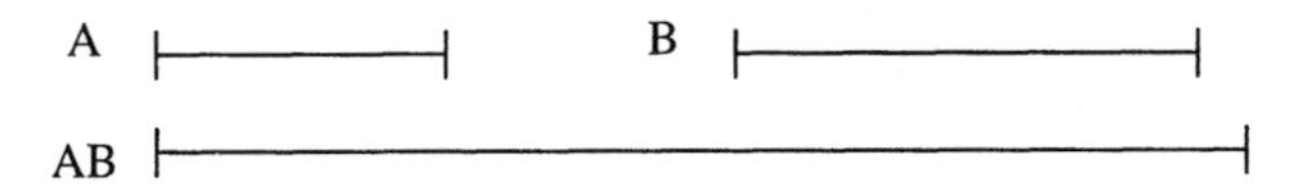

4. Let A and B be two natural numbers. Cut off from A, as many segments equal to B as possible. In case, this process *exactly exhausts* A as below

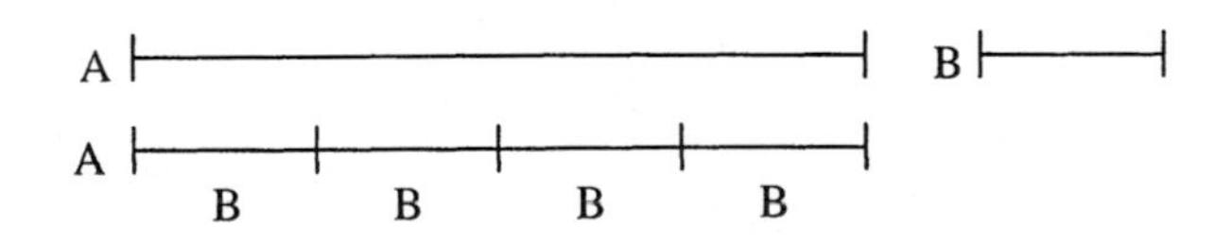

we say: *B measures A evenly.*

In case the process ends before A is exhausted, the part XY left to be covered being less than B, as below,

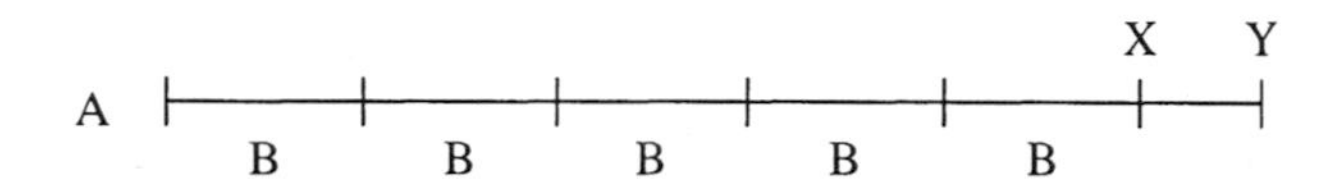

we say: B measures A not evenly and leaves a *remainder XY*.

It will be noticed that Euclid uses the term *measure* in the sense with which we use the term *divide*.

5. If a natural number A is measured evenly only by two segments, namely a unit segment and A itself, number A is said to be *a prime number*. And if a number A is measured *also* by numbers other than a unit and A itself, the number is called a *composite number*.

   These are the same definitions as of today. Thus,

   $$2, 3, 5, 7, 11, 13, \ldots$$

   are the first few primes and

   $$4, 6, 8, 9, 10, 12, \ldots$$

   are the first few composite numbers.

6. Let A and B be two natural numbers and let natural numbers

   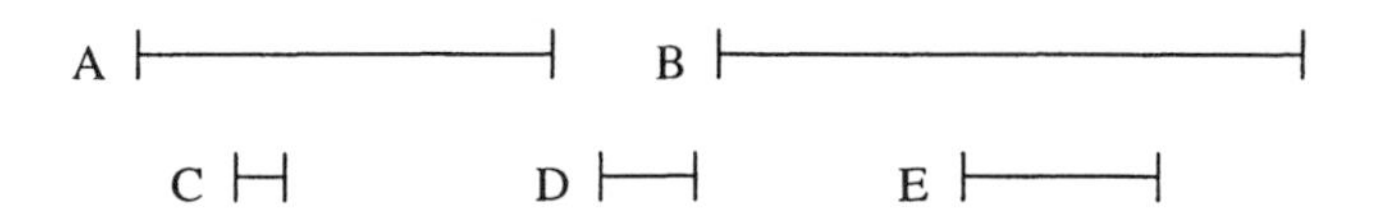

   C, D, ... measure both A and B. Then the greatest of the common measures C, D, ... is called the *greatest common measure* of A and B.

Thus if A and B are numbers 12 and 30, their common measures are

2, 3 and 6

and 6 is the greatest common measure of 12 and 30.

On the other hand, natural numbers A and B which have unit as the *only* common measure are said to be *relatively prime.*

Thus, 12 and 35 are measured respectively by 2, 3, 4, 6, 12 and 5, 7, 35; but they have no common measure. 12 and 35 are therefore, relatively prime.

8. Some number-theoretic results in the *Elements*.

Some preliminary notions given by Euclid at the start of Book VII are given above in Section 7. After stating his definitions at the start, he now uses them to give geometrical proofs of number-theoretic propositions. In Books VII, VIII and IX together he has given 102 such propositions. We shall give here only those which have held their place in number-theory even today. Some of them are the following:

1. The Euclidean Algorithm (Propositions VII.1 and VII.2).
2. Theorem: If a prime p divides a composite number $a \times b$, then it divides at least one of a and b (VII.30).
3. Theorem: Every composite number has at least one prime divisor (VII.31).
4. Theorem: Unique factorization theorem (IX.14).
5. Theorem: The number of primes is infinite (IX.20)

9. Euclid's proofs.

Euclid wrote the *Elements* about 2300 years ago, but has given for those number-theoretic theorems the same proofs as we now give. Only his language, terminology and style are different. The proofs or explanations of these results given below are essentially Euclid's; only the language and terminology are what are used currently. In the case of three of them namely VII.31, IX.14 and IX.20, we have given Euclid's proofs so that one could get a glimpse of Euclid's style.

1. His first result established in VII.1 and VII.2 is an algorithm which gives the greatest common divisor (g.c.d.) of two given natural numbers. In propositions VII.1 and VII.2, Euclid *proves* that the algorithm concerned gives the g.c.d. unfailingly. We shall only give the algorithm which emerges from his proof.

   The algorithm starts with applying the division algorithm to the division of one number by the other; followed by applying the division algorithm to the division of the second number by the remainder of the earlier step; and by repeating this process till the step in which the remainder is 0. Then the remainder of the step before this last step is the g.c.d. sought.

   Thus the Euclidean Algorithm consisting of all these divisions, when applied to the finding of the

   g.c.d. of 4669 and 1729

   would run as under.

$$\begin{aligned} 4669 &= 1729 \times 2 + 1211 \\ 1729 &= 1211 \times 1 + 518 \\ 1211 &= 518 \times 2 + 175 \\ 518 &= 175 \times 2 + 168 \\ 175 &= 168 \times 1 + 7 \\ 168 &= 7 \times 24 + 0 \end{aligned}$$

   Since the remainder at this step is 0, the remainder 7 of the previous step is the g.c.d. of 4669 and 1729.

   In VII.2, Euclid proves that the remainder at the penultimate step is a *common* divisor and also shows that it is the *greatest* of such common divisors.

2. The next important result proved by him in VII.30 is the theorem that

> if a prime p divides a composite number, then p divides at least one of the factors of the composite number whichever way we factor it.

Consider for example the *prime* number 7 and the *composite* number 210 which it divides. We can factor 210 in the following various ways:

$$\begin{array}{lll} 2 \times 105, & 2 \times 3 \times 35, & 2 \times 3 \times 5 \times 7 \\ 3 \times 70, & 3 \times 2 \times 35, & 3 \times 2 \times 5 \times 7 \\ 5 \times 42, & 5 \times 2 \times 21, & 5 \times 2 \times 3 \times 7 \\ 7 \times 30, & 7 \times 2 \times 15, & 7 \times 2 \times 3 \times 5 \end{array}$$

and it can be checked that the prime number 7 which divides 210 divides at least one of the factors of each of the twelve resolutions of 210 into factors.

On the other hand consider the case of 6 which divides the composite number 24. Now 24 can be factorized as 3×8, and we find that 6 which divides 24 does not divide any one of 3 and 8, the two factors of 24. This should not surprise us because 6 is not a prime. Consider now number 4 which divides 24. 4 also divides a factor of 24 whichever way we factorize it. Thus if p is is a divisor of a composite number c, p may or may not divide a factor of c when we factorize c. But if p is a prime then p certainly divides at least one factor of c in each of the factorizations of c into two or more factors.

3. The next result is the theorem in VII.31 that

> Every composite number has at least one prime divisor.

Here we give this theorem and its demonstration as in *Elements*.

Proposition 31.

Any composite number is measured by some prime number.

Let A be a composite number. I say that A is measured by some prime number. For, since A is composite, some number

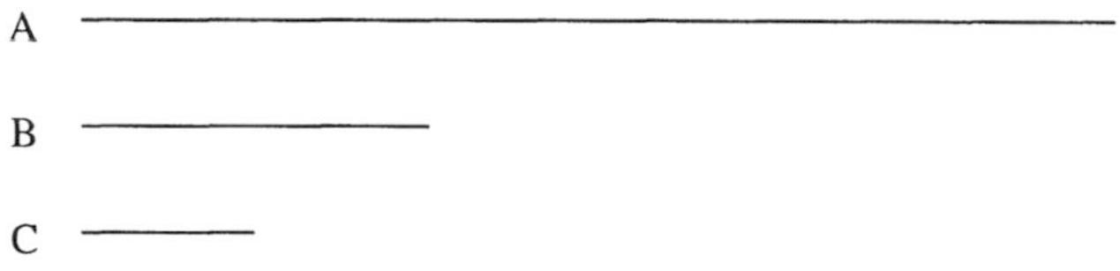

A

B

C

will measure it. Let a number measure it and let it be B. Now if B is prime, what was enjoined will have been done. But if it is composite, some number will measure it. Let a number measure it and let it be C.
Then since C measures B, and B measures A, therefore C alo measures A.
And if C is a prime, what was enjoined will have been done. But if it is composite, some number will measure it.

Thus if the investigation be continued in this way, some prime number will be found which will measure the number before it, which will also measure A.

For, if it is not found, an infinite series of numbers will measure the number A, each of which is the less than the other; which is impossible in numbers.

Therefore some prime number will be found which will measure the one before it, which will also measure A.

Therefore any composite number is measured by some prime number. **Q.E.D.**

4. The next two propositions which we have chosen to illustrate the contribution of Euclid to Number-Theory are propositions IX.14 and IX.20 from Book IX.

Let us have another look at the way Euclid argues, by observing his proof of Proposition 14 in his own words.

Proposition 14.

If a number be the least that is measured by prime numbers, it will not be measured by any other prime number except those originally measuring it.

For let the number A be the least that is measured by the prime numbers B, C, D;

I say that A will not be measured by any other prime number except B, C, D. For, if possible, let it be measured by the prime number E, and let E not be the same with any of the numbers B, C, D.

A B
E C
F D

Now, since E measures A, let it measure it according to F;

therefore E by multiplying F has made A.

And A is measured by the prime numbers B, C, D.

But, if two numbers by multiplying one another make some number, and any prime number measures the product, it will also measure one of the original numbers; [VII.30]
therefore B, C, D will measure one of the numbers E, F.

Now they will not measure E;
for E is prime and not the same with any of the numbers B, C, D.

Therefore they will measure F, which is less than A;
which is impossible, for A is by hypothesis the least number measured by B, C, D.

Therefore no prime number will measure A except B, C, D.

Q.E.D.

5. And now we come to the last of Euclid's contributions to number theory – the last of what we have selected as specimen of Euclid's ingenuity of discovery of results as well as the ingenuity he exhibits in the discovery of clever proofs of his theorems. The theorem which is the subject of proposition IX.20 as well as its proof have been acclaimed as the climax of Euclid's great theorems. The theorem asserts that

The number of primes is infinite.

The transparent simplicity and beauty of his proof of this claim is best seen in the following piece which is small but contains the essence of his argument.

Suppose B is the only prime. Then consider the number

$$A = B+1.$$

This number A must be composite, since B is the only prime and B is not the same as B+1.
Being composite, there must exist a prime which divides it; but B is the only prime and B does not divide B+1.

Thus, if B is the only prime, B+1 is neither composite nor prime. The contradiction proves that the claim "B is the only prime" is false.

Therefore, there must be at least one more prime.

Euclid's argument really ends here. But the procedure does not. How does one construct what should correspond to A above if the claim that there are only two primes is to be shown as a false claim? Euclid does not want to leave this to the readers and disturb them. He therefore starts with a claim that there are only three primes B, C, D and argues about whether

$$A = B \times C \times D + 1$$

is composite or prime. Having shown that this also leads to a contradiction, he shows that the claim that there are only three primes is false and claims that there should be at least one more. And so the argument proceeds automatically till he draws his final conclusion.

Let us see his proposition and his proof of it, all as it is in his book.

Proposition 20.

Prime numbers are more than any assigned multitude of prime numbers.

Let A, B, C be the assigned prime numbers;
I say that there are more prime numbers than A, B, C.

For, let the least number measured by A, B, C be taken, and let it be DE;
Let the unit DF be added to DE. Then EF is either prime or not.

First let it be prime;
then the prime numbers A, B, C, EF have been found which are more than A, B, C.

Next let EF not be a prime;
therefore it is measured by some prime number. [VII.31]

Let it be measured by the prime number G.

I say that G is not the same with any of the numbers A, B, C.

For, if possible, let it be so.

Now A, B, C measure DE;

therefore G also will also measure DE.

But it also measures EF. Therefore G, being a number, will measure the remainder, the unit DF;

which is absurd.

Therefore G is not the same with any of the numbers A, B, C.

And by hypothesis it is a prime.

Therefore prime numbers A, B, C, G have been found which are more than the assigned multitude of A, B, C.

Q.E.D.

10. Greek mathematics after Euclid.

The *Elements* of Euclid was the principal part of what Greece gave to the mathematical culture of the world. It takes notice of most of what was done by Euclid and the other geometers before Euclid. After Euclid, there were other good geometers who contributed substantial material to the total store of Greek geometry. But much of this work was isolated, separated from each other both in time and location.

However, the work of three of them, who came after Euclid deserves serious notice in the context of what Greece has left behind. The three mathematicians are Apollonius, Diophantus and the great Archimedes. The work of Archimedes and its importance to later European mathematics is considered in Chapter three of this book where it is assessed in its proper perspective. The work of Apollonius and Diophantus will be reviewed here.

11. Conic sections of Apollonius.

Apollonius, born around 225 B.C. was the greatest Greek geometer after Euclid. He studied in Alexandria, where Euclid taught about hundred years earllier. Apollonius worked on conic sections and composed a comprehensive and systematic treatise on the same. Even, the names *ellipse*, *parabola* and *hyperbola* for the three curves were first given by Apollonius and continue till today.

His *Treatise on Conic Sections* originally contained eight books. But only seven of them have been recovered, the first four in Greek and the next three in Arabic translations. In these seven books, a total of 147 propositions are established.

There are two important aspects in which the treatment of the subject by Apollonius differs from the one which we currently give to the subject. The first of these is in respect of the definitions of the curves. He defines them as

curves which result when planes cut cones

and derives all known properties of the curves from this definition and the rather intriguing figures that result in showing the curves as sections of a cone. In our present analytic treatment of the curves, we define them as loci of points whose distance s from a fixed line, *directrix*, bear a certain proportion, *eccentricity*, to their distances from a fixed point, *focus*. The terms, directrix, eccentricity and focus, are entirely absent in the books of Apollonius. Neither does he use the simple focus-directrix properties for his definitions nor does he prove these elegant properties anywhere in his books.

The second aspect in which Apollonius follows a treatment which differs from our present treatment is in his

concept of the normal.

He does not define a normal at a point on the curve as a line through the point perpendicular to the tangent at the point, nor does he prove this property. He defines the normal to the curve from a point P not on the curve as the line of the

shortest distance from P to the curve.

In fact, of the seven books of the treatise, one whole book, namely, *Book V* carries the title

normals as maxima and minima.

This book contains seventy-seven propositions, many of them about normals, some about construction of normals from given points, and yet some others about distances between related points which are either maximum or minimum. That Apollonius should devote one whole book out of his eight to this topic shows clearly the importance which Apollonius paid to this concept of maxima and minima which was, any way, a new concept in the study of curves. Those who followed him about fifteen hundred years later in Europe must have been duly impressed by the importance which Apollonius gave to this exciting aspect of the curves. They also, without fail, included this aspect in their studies of curves. The first contribution, for example, of Fermat to the method of differential calculus was the solution of a problem of maximum.

12. The *Arithmetica* of Diophantus.

Amongst Greek mathematicians, dominated by geometers, only one could be singled out for pursuing an altogether new direction. The person concerned was Diophantus – usually referred to as Diophantus of Alexandria on account of lack of any reliable personal information about him. He had a remarkable genius for constructing as well as solving problems involving numbers. Almost single-handedly, he created a new area in mathematics. This attracted the notice of many excellent European mathematicians of later centuries and stirred them to their very depths. Though Diophantus created very little of orthodox number-theory, the problems about numbers which he proposed and solved was a rich source of ideas and inspiration for those like Vieta, Fermat and Euler who later laid the foundation of number-theory.

Very little of the work of Diophantus has been recovered, the best known of which is

Arithmetica.

In this work, consisting of six *books*, Diophantus has collected about 132 problems which he had himself constructed. He has solved most of them. The problems cover varying areas. Some lead to determinate equations, some to indeterminate equations and yet some others to the determination of sides of right-angled triangles when various relations hold between them. The least that can be claimed for them is that they are, one and all, altogether exciting.

The following selection from his *Arithmetica*, amply illustrates the genius of Diophantus.

Example 1: Problem 17, *Book I.*

> Find four numbers, the sums in threes of which are given.

Since the solution required is in *positive rational* numbers, he lays it down as a necessary condition that one third of the sum of the four must be greater than any single one of the four sums in threes. He solves the problem when the four sums in threes were 22, 24, 27, 20. Since the solution is simple, it need not be given here. In fact, all problems included in *Book I* are easy. Consider another:

Example 2: Problem 30, *Book I.*

> Find two numbers if their difference and product are given.

Though simple, let us see his solution of the problem which he solves when difference is 4 and product 96.

Since the difference is 4, he takes the numbers to be

$$x + 2, \quad x - 2.$$

Their product is 96. Therefore

$$x^2 - 4 = 96$$

leading to $x = 10$ and the numbers are

$$12 \text{ and } 8.$$

To us all these problems in Book I are simple. We have a certain algebraic terminology and well-established procedures at our disposal. But at the time of Diophantus, the problems must not have appeared so simple and easy.

Problems in Book II onwards are hardly of such simplicity. The following are selections from these books.

Example 3: Problem 11, *Book II.*

> Given two numbers find a third which gives a square when added to either.

His solution is the following:

Given numbers: 2 and 3. Required number: x such that $x + 2$ and $x + 3$ are both squares.
To find x:

> Take the difference between the two given numbers; Here it is $3 - 2 = 1$. Resolve this difference, 1, into two factors: say 4 and $1/4$.
> Now take the square of half the difference, $4 - (1/4)$ between these factors.
> Since, $4 - (1/4) = 15/4$, the square of its half, $15/8$ is
>
> $$\frac{225}{64.}$$
>
> Equate it to $2 + x$, the lesser of the two numbers $2 + x$ and $3 + x$, required to be squares; and find x. Here,
>
> $$2 + x = \frac{225}{64}$$
>
> gives
>
> $$x = 97/64.$$
>
> Added to 3, it gives $3 + x = 289/64 = (17/8)^2$.
>
> Thus, $x = 97/64$ is the required number.

Problem 10, *Book III* is a kind of opposite of the above problem. It is the following:

Example 4: Problem 10, *Book III.*

> Given a number, find three others such that the product of every two of them gives a square when added to the given number.

He has solved it when given number is 12.

> Take *any* square, say 25; subtract 12 from it to get 13. Take 13 as the product of the first pair of numbers. And let the first two numbers be $13x$ and $1/x$.
> Again subtract 12 from any other square, say 16, and get 4. Take this as the product of the second and the third numbers. Since the second is $1/x$, the third number would be $4x$.
> Now, the third condition requires that
>
> $$4x \times 13x + 12$$
>
> be a square. This requires that
>
> $$52x^2 + 12$$
>
> to be a square; which it is if $x = 1$. And with this value of x, the three required numbers are
>
> 13, 1 and 4.
>
> The above solution is a little tentative. If choice of 25 and 16 would not have yielded a solution, the choice of these arbitrarily chosen squares would have had to be changed. And the answer also then would be different. For example, if 49/4 and 16 were chosen as squares in place of 25 and 16, the required numbers would be
>
> 11/8, 2/11 and 22.

With every subsequent book, the nature of the problems as well as the level of their difficulty changes. Consider, for example, the following three problems from Books IV, V and VI.

Problem 36, *Book IV*:

> Find three numbers such that the product of any two of them bears to their sum a given ratio.

Diophantus solves this problem by taking the ratios as 3, 4, 5 and gets 360/51, 120/23, 480/28 as the required numbers.

Problem 6, *Book V*:

> Find three numbers such that each minus 2 is a square and the product of any two minus their sum or minus the remaining number is a square.

Diophantus has solved this problem and has found 59/25, 114/25, 246/25 as the required numbers.

Lastly consider

Problem 13, *Book VI*:

> Find a right angled triangle such that its area minus either perpendicular gives a square.

Diophantus has found

$$12/5,\ 16/5,\ 4$$

as the sides of the triangle.

We shall close this section on Diophantus by referring to a strange but significant remark he made about rational numbers. In his solution of Problem 16 of Book V, he remarks that with rational numbers

> a difference of two cubes is also a sum of two cubes.

For example,

$$2^3 - 1^3 = 7 = \left(\frac{4}{3}\right)^3 + \left(\frac{5}{3}\right)^3,$$

as also

$$3^3 - 2^3 = 19 = \left(\frac{33}{35}\right)^3 + \left(\frac{92}{35}\right)^3.$$

How Diophantus has arrived at this result is not clear. He asserts it as a general truth but has not proved it.

There are other similar results. Their mysterious but interesting nature attracted the serious notice of later algebraists such as Vieta, Fermat and Euler. They studied such results in depth and contributed useful material to Number-Theory. Vieta, for example, has proved that

$$\text{if} \quad a^3 - b^3 = x^3 + y^3$$

$$\text{then} \;\; x = \frac{a(a^3 - 2b^3)}{a^3 + b^3} \;\; \text{and} \;\; y = \frac{b(2a^3 - b^3)}{a^3 + b^3}.$$

13. The greatest Greek gift.

In the above survey of the Greek work you have seen what the Greeks left behind: geometry as in the *Elements*, including Euclid's contributions to number-theory, geometry of the conic sections as in Apollonius, and number riddles of Diophantus – genuine challenges to the problem solving skills of those aspiring to be mathematicians.

But what could be termed the greatest gift to mathematical culture was the

Elements

of Euclid. The following is a thoughtful tribute which George Simmons pays to it.

> To mathematicians, some of the theorems in the *Elements* are important, some are interesting and some are both. However, the source of the immense influence of this book on all subsequent thought lies elsewhere, not so much in the exposition of particular facts as in the *methodology* of it all. It is clear that one of Euclid's main aims was to give a connected logical development of geometry in such a way that
>
> *every theorem is rigourously deduced from the "self-evident thruths" which are explicitely stated at the begining.*

> This pattern of thought was conceived by Pythagoras; but it was Euclid who worked it out in such stupefying detail that for more than 2000 years no one was capable of doubting that his success had been complete and final ...
>
> This happy state of affairs in geometry led to the hope that in a similar way the remotest truths of science and society could be discovered and proved by simply pointing out those things that are self-evident and then reasoning from these foundations. No more *attractive* or *tenacious* idea has ever appeared in the intellectual history of the Western world. The prestige of geometry was so great, especially in the seventeenth and eighteenth centuries, that *true knowledge* in any field almost required the Euclidean deductive form as a *seal of legitimacy.* The more disorderly branches of knowledge, which evaded this pattern were considered to be somehow less respectable, a stage or two beneath the aristocratic disciplines. ...
>
> All in all, for more than 2000 years the intellectual architecture of the *Elements* has rivalled the Parthenon as a symbol of the Greek genius. Both have deteriorated somewhat in recent centuries, but perhaps the book has sustained less damage than the building.

This was the real greatness of Euclid and this is the greatest legacy he left behind to give to mathematical propositions and their proofs the discipline which mathematics is proud to possess today.

Exercises

1. Use the Egyptian method of multiplication given in Sec. 3 and find

   (a) the product 29×21

   (b) the product 91×19

   (c) the square of 81.

2. (a) Use the Egyptian method (not just formula) and find the area of a circle of diameter 18.

   (b) Use formula πr^2 and obtain the same area using a calculator.

   (c) Compare the two values and obtain the resulting value of π.

3. (a) Construct a 6×2 rectangle.

   (b) Use the method given in Sec. 6 and square the rectangle of part (a).

4. (a) Construct a triangle of sides 5, 6, 7.

   (b) Construct a rectangle equal in area to the triangle of part (a).

   (c) Square the triangle in (a).

5. (a) Construct a trapezoid ABCD such that AB $= 6$ and angle ABC $= 90°$, BC $= 13$, angle BCD $= 90°$ and CD $= 4$.

   (b) Square the trapezoid of part (a).

6. Read the algorithm called Euclidean Algorithm given in Sec. 8 of the book and

   (a) find the g.c.d. of 385 and 2717

   (b) find the g.c.d. of 10659 and 8211

   (c) show that numbers 3553 and 805 are relatively prime.

7. Write 194040 as a product of primes.

8. BB′ is the minor axis of an ellipse, the lengths of whose semi-major and semi-minor axes are a and b. P is a point on BB′ such that BP $= a^2/b$. Show that no point on the ellipse has its distance from P larger than a^2/b.

9. Find four rational numbers, the sums of which taken three at a time are 22, 24, 27 and 20.

10. Find a rational number x such that $x + 3$ and $x + 7$ are both squares of rational numbers.

11. (a) Find a rational number x such that $x - 3$ and $x + 7$ are both squares of rational numbers.

    (b) Also find the square numbers which are equal to $x - 3$ and $x + 7$.

12. Solve Problem 36, *Book IV* of Diaphantus given on page 33.

Further Reading

1. Bunt, Jones, Bedient, *The Historical Roots of Elementry Mathematics*, Dover Publications, New York, 1988.

2. Sir T. L. Heath, *The Thirteen books of Euclid*, Dover Publications, New York, 1956.

3. Sir T. L. Heath, *Apollonius of Perga*, Barnes and Noble Inc., New York, 1961.

4. Sir T. L. Heath, *Diophantus of Alexandria*, Cambridge at the University Press, 1910.

5. George F. Simmons, *Calculus Gems*, McGraw-Hill, Inc., New York, 1992.

CHAPTER TWO

THE INDIAN ARITHMETIC
ITS ENTRY AND ACCEPTANCE IN EUROPE

1. Indian Numerals and Algorithms.

Greek mathematics, as we saw in the last Chapter concentrated on Geometry and almost on Geometry alone. Under them Geometry acquired such a high status that other work in mathematics, how so useful it may be, was just ignored and cast aside as something basic and not deserving serious attention of intellectuals.

They had a numeral system. But it was weak and not of much use. But since philosophers disdained it, no particular effort was made to replace it. This continued all the time the Greek mathematics was in its very zenith. And then, around the fourth century, the pursuit of mathematics started decaying in Greece and everywhere else in Europe.

The cultivation of mathematics got revived only in the early thirteenth century. It was not Athens or Rome or Alexandria which were then any great centers of culture or trade. Such a center was at Baghdad. European traders, particularly from Italy and the surrounding Mediterranean countries regularly commuted between their home countries and Baghdad. Such traders saw different cultures abroad and learnt from them. One such trading party headed by the father of Leonardo da Pisa, used to have the brilliant young son Leonardo, as a member. It was this young boy who was the first to notice that their Arabian counterparts used a numeral system for their calculations different from their own. He saw that the Arabian traders were appreciably quicker in their calculations and was convinced that they were in possession of a more efficient numeral system. When he pursued the matter, he was told that the numerals which they used at Baghdad were *Indian numerals* and the algorithms which they were employing in their calculations were *Indian Algorithms.*

Pleasantly struck by its efficiency and by its unquestionable superiority over the *Abacus* in use in his country, he decided to bring it to Europe and acquaint the intellectual community back at home with it. He studied it thoroughly and prepared a book on it. He made it public under the title *Liber Abbaci.*

2. Fibonacci and his *Liber Abbaci.*

> "These are the nine figures of the Indians
>
> 9, 8, 7, 6, 5, 4, 3, 2, 1
>
> With these nine figures and with this sign 0, which in Arabic is called Zephirum, any number can be written as will below be demonstrated".

This is the sentence with which opens the first chapter of *Liber Abbaci* of Fibonacci. It was this book of Fibonacci written in A.D. 1202 and revised in 1228 which was principally responsible for the entry into Europe of the Indian numerals, the Indian positional decimal system of writing numbers and the Indian algorithms for addition, subtraction, multiplication and division.

The real name of Fibonacci was Leonardo, Leonardo da Pisa. His father was a trader commuting between Italy, his native country and Baghdad, the trade center of the Islamic empire. Even in his early teens, Fibonacci accompanied his father on the latter's trade-trips and helped him in the purchase of merchandise and in preparing bills and making payments. He was surprised that the native Arabian traders prepared the bills appreciably faster than he could, on his Abacus, the computing device then in use in Europe. When his young and inquiring mind looked into it more carefully, he discovered that the Baghdad traders used a different numeral system and a different set of algorithms for the arithmetic operations of addition, subtraction, multiplication and division. On enquiry, he was told that that was the Indian decimal system of writing numbers with the help of nine figures 9, 8, 7, ..., 2, 1 and the sign 0, and a cleverly devised place-value system; and that the operational methods were the *Indian Algorithms* for the various arithmetic operations. He immediately took up the study of the system with the help of Arab experts and mastered it in no time. Impressed by the all-round superiority of the system over the one which then prevailed in Italy, he thought of bringing the same to the notice of the traders and the educated people of Europe; and for this purpose, wrote his famous treatise *Liber Abbaci.*

In spite of this work and effort, Fibonacci could not succeed in converting any very large number of either merchants or mathematicians to his view. The reason was that an operationally simpler system of computing already existed in Europe.

3. Abacus : The prevalent system in Europe.

The prevalent number system in Europe then was that of the Roman numerals, the digits in which were

I, II, III, IV, V, VI, ..., X, XI, ..., L, ..., C, ..., D, ..., M, ...

I for 1, V for 5, X for 10, L for 50, C for 100, D for 500, M for 1000, and so on.

Unlike the decimal system, where the places denote values in multiples of 10, in the Roman system, the multiples were alternately *five times* and *twice*. Thus V is five times I, X is twice V, L is five times X, C is twice L, D is five times C, M is twice D, etc. Thus as in the decimal system, there is in this system also, a regularity, different though from the former.

In the Roman numeral system, there was one important simplicity. In the decimal system definite places are reserved for digits indicating units, tens, hundreds, etc.; and if there be no digit in any particular place, the gap is shown by the sign 0. In the Roman system, there is no such reservation, except that the larger unit is to the left of the smaller one. This, occasionally, simplifies the writing. For example, the number *one thousand and one* is written as 1001 in one system and as just MI in the other. This does not mean that numbers written in Roman numerals are shorter in size than the same written in decimal system. The number *ninety eight* is 98 in one system and LXXXXVIII in the other.

The real difference lies in the fact that in the decimal system numbers are *written* and operations with them are done in *writ-*

ten forms. In the Roman system, numbers are *shown* by counters on the Abacus and operations are done by the *movement of the counters.*

4. Operations on the Abacus.

To understand how operations are done on the Abacus let us consider a small-sized Abacus consisting of four vertical parallel lines, as shown below, labeled M, C, X, I running from left to right and three dotted lines labeled D, L V midway between every two. Counters based on lines M, D, C, L, X, V, I represent respectively

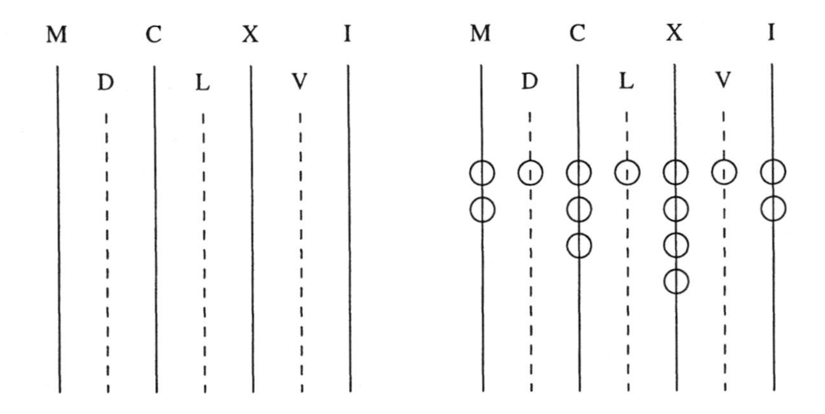

A miniature Abacus

Abacus with a number

the number of thousands, five hundreds, hundreds, fifties, tens, fives and ones. Thus on the right Abacus shown above, the number recorded is MMDCCCLXXXXVII, that is, the number 2897.

When doing additions and multiplications on the Abacus, as with decimal system numbers, occasions arise when counters from one line are *carried over* to the line on the left. Thus 5 counters on I are carried over into 1 counter on V; 2 counters on V are converted into 1 counter on X; 5 counters on X are carried over as 1 counter on L, etc.

When doing subtractions and divisions, if need be, we *borrow* counters from a line to the *left*. For example, if 4 counters on I are to be deducted from 2 counters on I, we borrow 1 counter from V and convert it into 5 counters on I, thus making a total of 7 counters on I from which the required 4 counters can be deducted. Similarly, if 1 counter is borrowed from X and brought to V, it becomes 2 counters on V; with similar working rules in respect of borrowing from other lines.

These *carryings-over* and *borrowings from* are not different from what we do in operations in the Decimal System. Thus in doing the addition

$$\begin{array}{rr} & 2\ \ 5 \\ + & 1\ \ 6 \end{array}$$

we get 11 as the sum of the two digits 5 and 6 in the unit's places. Of this sum, we retain 1 as the digit in the unit's place of the sum, and carry 10 over to the ten's place where as a digit it becomes 1. Adding this carried over 1 and the original 2 and 1 of the two numbers, we get 4 in the ten's place of the sum. The required sum is 41.

Similarly in doing the subtraction

$$\begin{array}{rr} & 2\ \ 5 \\ - & 1\ \ 6 \end{array}$$

since 6 is larger than 5, we borrow 1 from the ten's place of 25. After this, in the unit's place of the first number we get 15 from which we deduct 6 and get 9 in the unit's place of the result. After the borrowing, the first number has 1 in the ten's place from which the 1 of 16 is deducted to result in a 0 in the difference. Thus the final required difference is 09 or 9.

Procedures on the Abacus are exactly similar. We illustrate them with the same two examples.

EXAMPLE 1: To get the sum of XXV and XVI.

Enter numbers XXV and XVI on the Abacus, XVI below XXV as shown in the first figure below. Thus XXV + XVI = XXXXI.

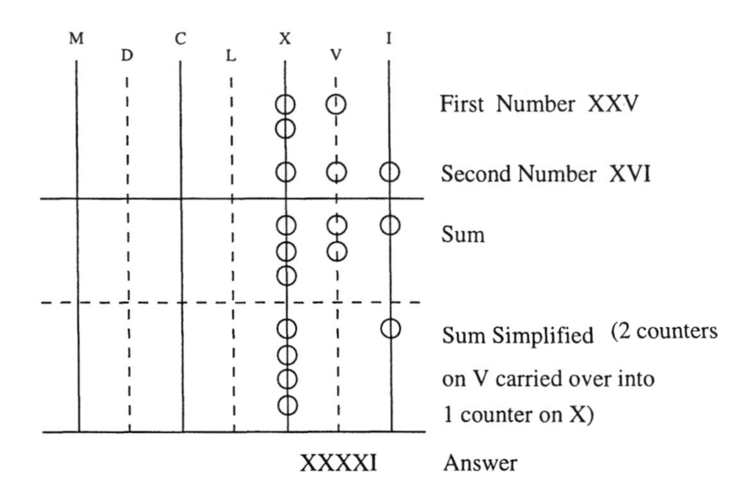

EXAMPLE 2: To get the difference XXV − XVI.

Enter the two numbers XXV and XVI on the Abacus, the second one under the first as shown below. Since there is 1 counter

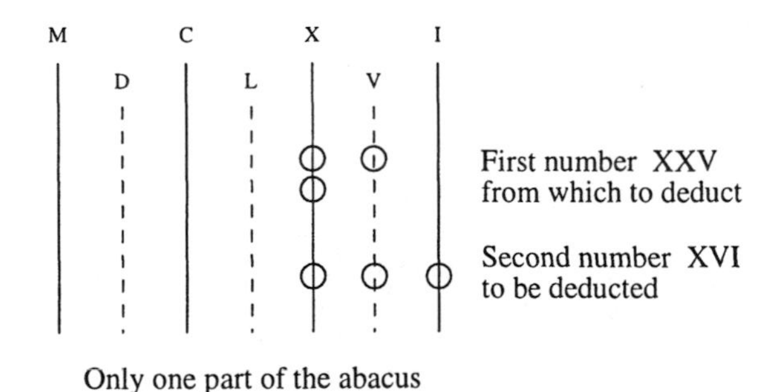

Only one part of the abacus

on I in the number (B) to be deducted and no counter on I in the number (A) from which we have to deduct it, we borrow 1 counter from V of A and get 5 counters on I of A. Now there remains no counter on V of A and there is 1 counter on V of B. We therefore borrow 1 counter from X of A and get 2 counters on V of A and leave only 1 counter of A on X. The position, now, after these borrowings, is as shown below.

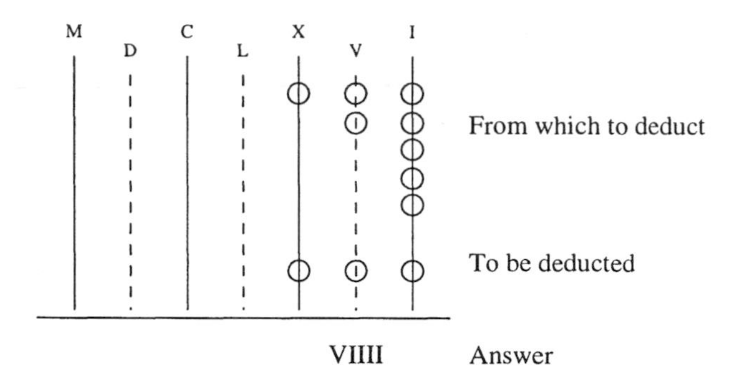

and the answer is now easily written.

We have shown the work elaborately. With practice, a lot of work is done orally and result obtained in only 2 steps.

5. Multiplication and Division on an Abacus.

Multiplication is a little more difficult. In the decimal system we use multiplication tables to facilitate the multiplying procedure. For the Abacus, we remember a list of rules according to which counters of the multiplicand move when they are multiplied by counters on I, V, X, L, etc. in the multiplier. We shall first give a short list of such rules covering multiplication by counters on I, X, C, and illustrate how they work.

Rule A1. Multiplication by a counter on I keeps all counters in the multiplicand in the *same* position.

Rule A2. Multiplication by a counter on X moves every counter in the multiplicand 2 places to the left.

Rule A3. Multiplication by a counter on C moves every counter in the multiplicand 4 places to the left.

Rule B. In all cases, if multiplication is by n counters on a line, the corresponding movement is to be repeated n times.

We shall illustrate these rules by an example.

EXAMPLE 3: Multiply XXXVII by CCXI.

Enter multiplicand XXXVII on the Abacus and write multiplier aside. Multiply by CC; that is, shift multiplicand 4 places to the

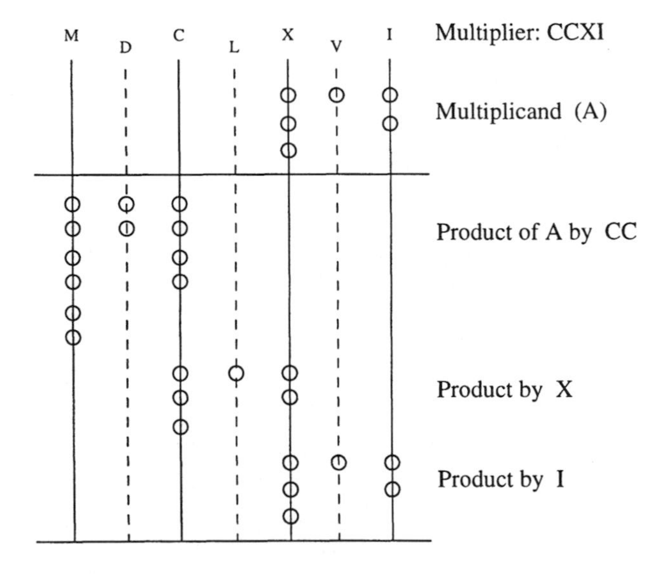

left and repeat. (Rules A3 and B).

Multiply by X; that is, shift multiplicand 2 places to the left.

Lastly, multiply by I; that is, write the multiplicand as it is. To get the answer, now, we have to add and simplify. The answer is MMMMMMMDCCCVII.

Rules regarding multiplication by counters on V, L, D are less simple. They are:

Rule C1. (a) Multiplication by a counter on V shifts counters on I, X, C one place to the left.

(b) Multiplication by a counter on V of counters on V, L, D results in 2 counters 1 place to the left and 1 counter in its old place.

Rule C2. (a) Multiplication by a counter on L shifts counters on I, X, C three places to the left.

(b) Multiplication by a counter on L of counters on V, L, D results in 2 counters three places to the left and 1 counter two places to the left.

Similar rules could be stated for other multiplications. But the rules stated are illustrative enough.

EXAMPLE 4: Multiply XVI by CLV.

Enter multiplicand XVI on Abacus, write multiplier CLV aside. Multiplications by C, L, V are done according to the rules given

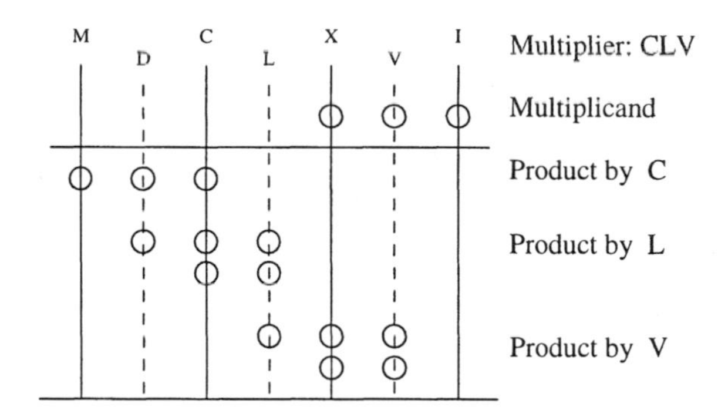

above. The products are shown on the Abacus. To get the result, we add and simplify.

The answer is MMCCCCLXXX.

Division is the reverse process. As in the decimal system, the figures in the quotient are fixed by reasonable guesses.

EXAMPLE 5: Divide DLXXVII by XVII.

In division on an Abacus, it is convenient to enter the divisor at the top and the dividend below it. At every stage, the divisor is multiplied by the part quotient and the product subtracted from the dividend. Steps below explain the procedure.

Enter the numbers on the Abacus as shown. The dividend A has one counter on D; therefore 3 counters coming on C is a reasonable guess. We choose XXX as the first part of the quotient. By this we multiply divisor B and deduct the product XXX times B from A and get A′ = LXVII.

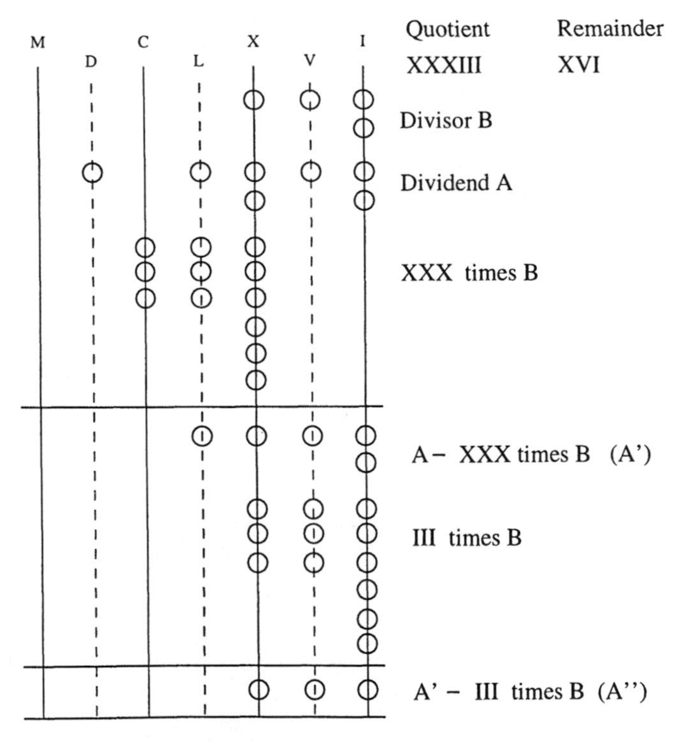

A'' is less than divisor B = XVII. Division is over.

A′ has 1 counter on L; therefore 3 counters on X is a safe guess;

we choose III as the second part of the quotient, and deduct III times B from A′ to get A″ = XVI.

A″, namely, XVI is less than the divisor B = XVII. Hence division is over; XVI is the remainder and XXXIII is the quotient.

6. Abacus versus the Decimal System.

The short demonstration above gives a fair idea of the strengths as well as the weaknesses of the Abacus.

To us, Abacus appears cumbersome and time-consuming. It is natural that it should so appear; it is new to us and we have not practiced it; instead, we have been constantly practicing an alternative, namely, the algorithms of the decimal system.

On the Abacus, the only mental activity involved is the moving of counters according to a certain set of simple rules. This requires very little prior preparation either to understand or to operate - an aspect which must have worked quite heavily against giving up the Abacus and accepting the decimal system in its place. Till about the twelfth century Europe knew Abacus as the only computing device and it had been working for so many hundreds of years and so very satisfactorily. Neither mathematicians nor merchants saw any great reason why they should give up what was firmly rooted in the public mind and in their own day-to-day work for something else, how so good that something else may be. The greatest argument in favor of Abacus was that *it existed* and the decimal system did not.

But the Abacus has its limitations. The simple Abacus used above for the purpose of illustrating the device has seven places to operate with, but cannot handle multiplications in which products exceed the limit of a few thousands, because there was no wire there beyond M for the thousands. If we have to work out multiplications whose products come to about a million, the Abacus would have to have at least thirteen wires. And a million was not a big enough number even for the mathematics or the trade-transactions of the twelfth century. With a larger Abacus, the number of rules would also be larger in number and more complex in nature for an average worker to remember with comfort and confidence.

Even granting that the decimal system and its algorithms have numerous advantages over the Abacus, to accept it and discard an existing system which was infinitely simpler than the new one, was not an easy decision to take. Proficiency in work with the algorithms of the decimal system required preparation extending over long periods. The schools today take a minimum of three to five years to train students in the system and bring them to an average working proficiency. Motivated adults may take less; but certainly not as little as they would, to pick up work on an Abacus. After all, counting and moving counters on wires is much simpler than learning to *write* numbers having queer shapes, remembering their names, remembering addition and multiplication tables and then applying them to varied operations as per the need.

The objection to accepting the decimal system was understandable from this point of view. What was not quite understandable was when this objection was at times vehement. Objections and counter-objections, charges and counter-charges were freely traded against each other. In fact the "algorists" (those in favor of the decimal system) and the "abacists" fell into two clear-cut camps. All kinds of fears and suspicions, right and wrong, were put forward. Even royal intervention was invoked to prevent the increasing inroads of the new system.

In 1299, the city of Florence issued an ordinance which prohibited the writing of numerals in columns, as well as the use of Indian numerals. The reason was supplied by a Vatican treatise on book-keeping: it is so easy, the treatise said, to change a 0 to a 6 or a 9. It was not so easy to falsify Roman numerals. Even in 1594 an Antwerp cannon warns merchants "not to use the new numerals in contracts or in drafts." When in the fifteenth century, the Indian numerals reached Germany, the reception was not exactly smooth. In 1494, the Mayor of Frankfurt cautioned the clerks to use these numerals sparingly. In this later part of the fifteenth century, German textbooks on Arithmetic advocated teaching "calculations on lines and with numerals." That they should prefer to advocate the use of both the Abacus and the Indian numerals is a clear indication of the prevalent mode of the general public which was still wavering in the matter of accepting the new numerals outright.

It took a few centuries for the issue to be finally settled. The overwhelming advantages of the decimal system and the extremely restrictive limitations of the Abacus for trade as well as mathematics gradually became more and more clear. More and more were converted and the number of abacists dwindled down to a negligible minority. Finally, the Indian numerals and Indian algorithms were accepted and accepted for good.

Today, different countries speak different languages, use different alphabets, have different food habits and different customs, have different social values and have different political aspirations; *but almost all nations of the world use the same numerals and the same algorithms* - the Indian numerals and the Indian algorithms.

7. Other contributions from India.

The development of the positional decimal system was one of the three most influential contributions of India to the development of mathematics. The other two were elements of Algebra and the introduction of the *sine function* in Trigonometry which replaced the inconvenient Greek table of chords.

The algebra which ultimately entered Europe was essentially the *Bijaganita* of the Indian mathematicians *Brahmagupta* and *Bhaskaracharya*. Brahmagupta's contribution to algebra was indeed of a high order. Here we find the general solution of the quadratic equation: and it is remarkable that he mentions both the roots even when one of them is negative. The systematized arithmetic of negative numbers is first found in his work in the following words:

> "Positive divided by positive or negative by negative is affirmative. Cipher divided by cipher is nought. Positive divided by negative is negative. Negative divided by affirmative is negative. Positive or negative divided by cipher is a fraction with that for the denominator."

Though those parts of the above paragraph which involve cipher, i.e., zero are incorrect, the rest is correct and shows that negative numbers were acceptable to him as much as positive numbers. It should also be remarked that the Indian mathematicians, unlike the Greeks, regarded irrational roots of numbers as natural and fully acceptable numbers. Whether this was, as some remark, a matter of "logical innocence rather than of mathematical insight", it was, any way, a great step and was of enormous help in the development of algebra.

The position in respect of division by zero was quite correctly put by the eleventh century Hindu mathematician Bhaskaracharya in the following words:

> "Dividend 3. Divisor 0. Quotient, the fraction 3/0. This fraction of which the denominator is cipher, is termed an *infinite quantity.* In this quantity consisting of that which has cipher for a divisor, there is *no alteration*, though many may be inserted or extracted."

Indian work in the subject of *indeterminate equations* was also commendable. It was Aryabhat who first proposed (AD 499) a problem leading to an indeterminate equation of the first degree in two variables and solved it. Brahmagupta (AD 528) even gave a general solution of the equation $ax + by = c$ in the form

$$x = p + mb, \quad y = q - ma$$

also showing awareness that $ax+by = c$ has integral solutions if and only if a and b are relatively prime. Brahmagupta also proposed a problem leading to an indeterminate equation $x^2 = 1 + py^2$ and solved the same if p had the special values ± 1, ± 2 and ± 4. Bhaskar (AD 1150) later solved the indeterminate equation $x^2 = 1 + py^2$ for all values of p by a method he called the *Chakrawal* (cyclic) method. It is remarkable that he solved the equation $x^2 = 1+61y^2$ and obtained the solution x = 1,776,319049 and y = 22,615,390.

But this work of Indian mathematicians did not reach Europe in good time to influence the main European stream of mathematics. Lagrange and others solved the problems independently of the Indian work.

In Trigonometry, as was remarked above, the contribution of India was the introduction of the *sine function* and the construction of a table of sines of angles between 0 and 90 degrees for 24 equally spaced angles, 0°, 3° 45′, 7° 30′, ..., 90°. In the extant Sanskrit literature on the subject, we find this table and the following method to construct it in the small treatise *Aryabhatiya* (AD

499) of Aryabhat. Aryabhat expresses arc-length and sine-length in terms of the same unit, to obtain which he takes the radius to be 3438 and circumference to be $360 \times 60 = 21600$ units, so that the arc-length of a quadrant from 0° to 90° becomes 5400 units. The first two entries in his table of sines are:

$$s_0 = \sin 0° = 0$$

and $s_1 = \sin 3°\ 45' = \text{arc-length of } 3°\ 45' = 5400/24 = 225.$

For getting the sines of the next successive multiples of 3° 45′, he uses the *recurrence formula*

$$s_{n+1} = s_n + s_1 - S_n/s_1$$

where, S_n denotes the sum $s_0 + s_1 + \ldots + s_n$. Thus since $s_0 = 0$ and $s_1 = 225$,

$$\begin{aligned} s_2 &= s_1 + s_1 - (s_0 + s_1)/s_1 \\ &= 225 + 225 - 1 \\ &= 449 \end{aligned}$$

and

$$\begin{aligned} s_3 &= s_2 + s_1 - (s_0 + s_1 + s_2)/s_1 \\ &= 449 + 225 - (0 + 225 + 449)/225 = 674 - 2.99 \\ &= 671 \end{aligned}$$

and so on, till $s_{24} = \sin 90° = 3438.$

8. Entry into Arabia and thence into Europe.

All this would have remained for long a closed book to the outside world had not Arab travelers and scholars, either on their own or at the instance of their Caliphas at Baghdad visited Indian centers of learning, met the masters, studied under experts, mastered the knowledge, taken it back to Baghdad, added a number of simplifications, modifications, innovations, illustrations and applications and published it in simple running Arabic prose, and carried the same to Europe or helped European traders and scholars to do it and facilitated its entry into the main stream of European mathematics.

This was indeed invaluably good work; and Mathematics must owe a great debt to Arab scholars for this, so must Greece, and so must India also.

Exercises

1. (a) One simple method to multiply two numbers is the following. (This method originated in India.)
Suppose the two numbers are one of 3 and the other of 4 digits. Construct a blank rectangle 3 by 4 and divide into 12 squares as in Fig. 1. Enter the 4-digit number at

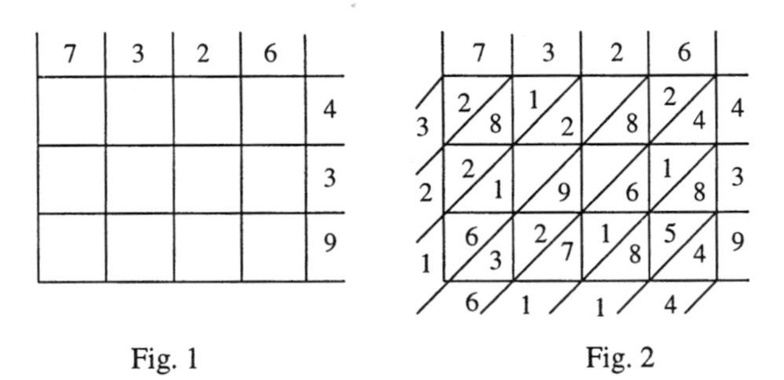

Fig. 1 Fig. 2

the head of the columns; the three digit number at the right of the rows as in Fig. 1. Now divide each square by means of the diagonal. Multiply digit by digit and enter the 12 products in the squares, placing the figure in the unit's place below and the digit in the ten's place above the diagonal of the square common to the column and the row of the digits whose product is entered. Now add diagonal-wise. Read the sum from the left side. That is the product: it is 3216114 here.

(b) Use this method and obtain the products

$$739 \times 983, 4603 \times 27 \text{ and } 10056 \times 73.$$

2. Draw a miniature Abacus of 7 wires. On it show the numbers
   (a) DLXVI (b) CXXXI

3. Convert (a) 1993 and (b) 1001 into Roman numerals and enter them on an Abacus.

4. (a) Find the sum of numbers (a) and (b) of Ex. 2.
   (b) Find the difference DLXVI – CXXXI.
   (c) Find the difference 1993 – 1001 in the decimal system and also on the Abacus.

5. Use an Abacus and
   (a) Find the product LXII and XXII.
   (b) Find the square of XXII.
   (c) Find the cube of XII.

6. (a) Read carefully how Aryabhat constructed his table of sines and use the formula

$$s_{n+1} = s_n + s_1 - \frac{s_n}{s_1}$$

   and compute s_4 and s_6.
   (b) Divide s_4 and s_6 by 3438 and verify how close Aryabhat's values of $\sin 15°$ (s_4) and $\sin 22°30'$ (s_6). Compare with the values of the same as available on a calculator or a modern table of sines.

Further Reading

1. Colebrook H. T., *Algebra With Arithmetic and Mensuration From the Sanskrit of Brahmagupta and Bhascar*, 1817.

2. Howard Eves, *The Great Moments of Mathematics – Before 1850*, Dolciani Series, Mathematical Association of America, New York, 1985.

3. J. F. Scott, *A History of Mathematics*, Taylor and Francis Ltd., London, 1969.

4. C. B. Boyer, *History of Mathematics*, John Wiley and Sons, Inc., New York, 1968.

CHAPTER THREE

THE METHOD OF INTEGRAL CALCULUS

1. The Greek roots of the method.

This is a chapter on the method of Integral Calculus. The *roots of integral calculus are very much in the work* of a great Greek mathematician-Physicist,

Archimedes

one of the three all-time greats in the field of Mathematics and Physics, the other two being Newton and Gauss.

Archimedes wrote two outstanding treatises,

The measurement of a circle
and *The quadrature of the parabola*

bearing on the subject.

The first is on an expression for the area of a circle. The second is on the problem of the area of a parabolic segment bounded by

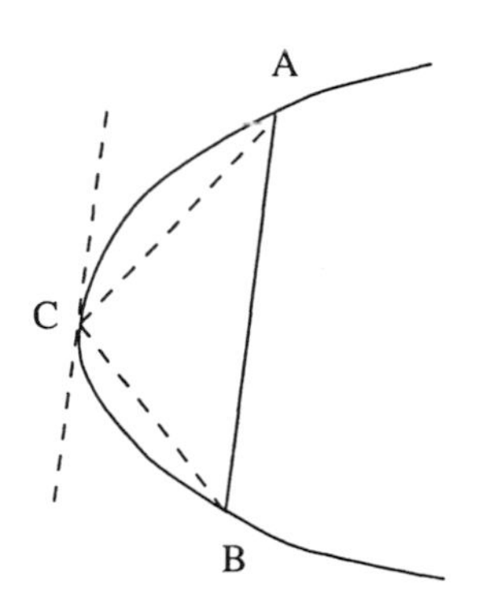

an arbitrary chord AB and the parabolic arc ACB. Archimedes showed that it is equal to four-thirds of the area of triangle ACB, where C is the point on the parabolic arc where the tangent is parallel to chord AB.

The method used in both the cases, that of a circle and a parabolic segment, was the method of successive approximation. For the circle he used inscribed regular polygons for the successive approximations. For the parabolic segment he used a succession of inscribed triangles to get a similar result. The principle he used in both the cases was that the error committed in these approximations can be *exhausted*.

We shall see how he did it. It is important to note because it has a direct link with integral calculus as was developed later after about fifteen centuries.

2. Measurement of the circle.

Proposition 6 in Book IV of the *Elements* was the starting point of the work of Archimedes in the matter of measurement of the circle. In this proposition, Euclid shows that a square ABCD can be

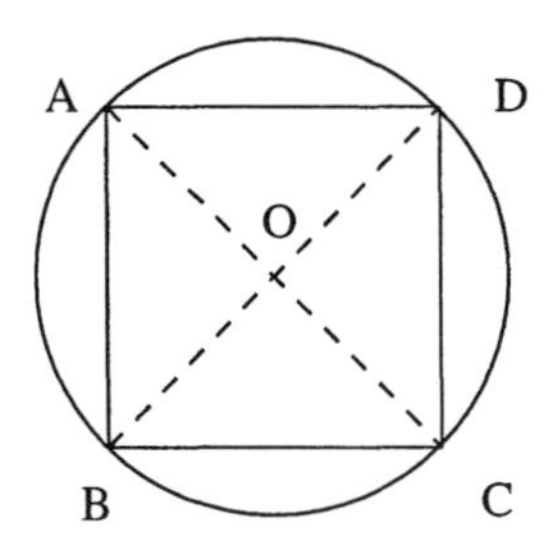

inscribed in a circle, by first drawing two perpendicular diameters AC and BD and then joining A, B, C, D to obtain the square.

Archimedes made two uses of this construction. From the center O of the circle, he drew lines OE, OF, OG and OH perpendicular to AB, BE, CD and DA to meet the circle in E, F, G and H. Joining the 8 points A, F, B, ..., H, A he obtained an inscribed 8-sided polygon which one can easily prove to be regular. Using the same

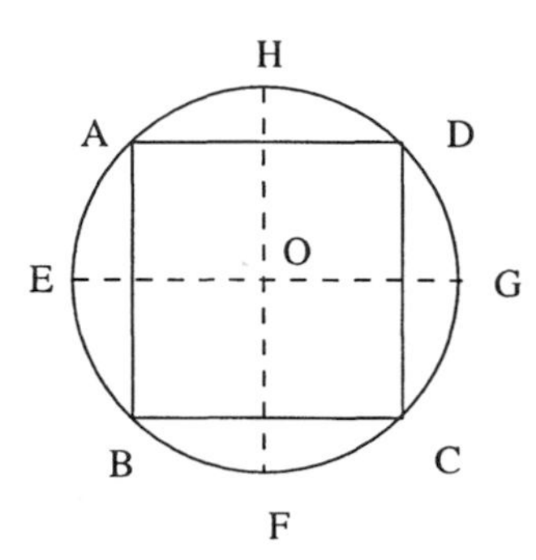

technique once again, he obtained an inscribed regular 16-sided polygon. By repeated application of the technique he showed that it is possible to inscribe in a circle a regular polygon of 2^n sides for every value of n.

The second way in which he used Euclid's proposition was the following. He considered the *area* of the square as a first approximation to the *area* of the circle, the area of the 8-sided polygon as a second approximation , the area of 16-sided polygon as a third approximation and so on. He observed that the first approximation was very rough, the next successive approximations were closer and closer. In the first approximation, the *error* was four times the area of the region AEB. In the second approximation, the region of the triangle AEB which was a part of the error is now a part of the 8-sided polygon and the corresponding error part is now the sum of areas APE and EQB. Thus the error becomes smaller in this case by as much as the area of the triangular region AEB. Obviously the argument can be applied to the 16-sided polygon and other successive regular polygons of 2^n sides inscribed in the circle and it could be shown that the error decreases with each new approximation. What Archimedes demonstrated amounts to the following.

> The error in approximating the area of the circle by the area of a regular inscribed polygon of 2^n sides can be *exhausted* in the following sense:
> If E is any preassigned positive number, there exists an

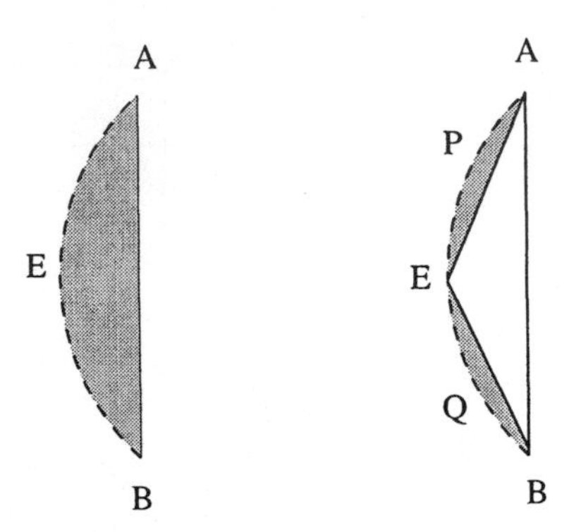

inscribed polygon of 2^n sides, such that

$$\text{Area (Circle)} - \text{Area (Polygon)} < E.$$

Archimedes next extended this result to *any* inscribed regular polygon and the final conclusion now became the following.

> The error in approximating the area of a circle by the area of a regular polygon inscribed in the circle can be *exhausted* in the same sense as above.

In the second extension he showed that a similar result is true for regular polygons circumscribing the circle. The only difference

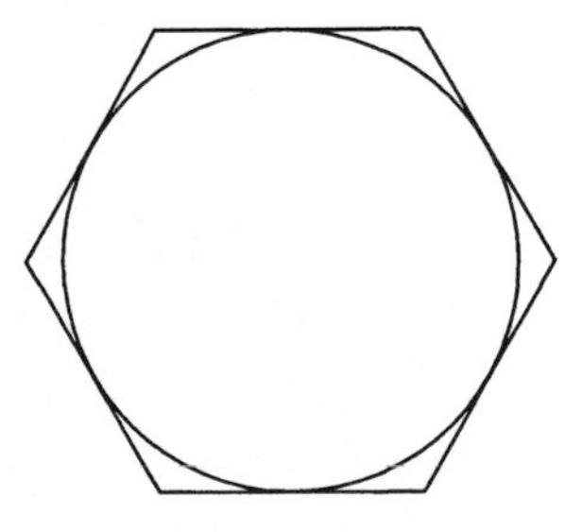

in this and the earlier case is that if one approximates the area of a circle by the area of a circumscribed polygon, the error is by excess, the area of the polygon being larger than the area of the circle. Otherwise, the result is the same and can be stated in full as follows:

> The error in approximating the area of a circle by the area of a circumscribed regular polygon can be *exhausted* in the same sense as above; namely
>
> If E is any preassigned positive number, there exists a circumscribed regular polygon such that

$$\text{Area (Polygon)} - \text{Area (Circle)} < \text{E}.$$

For taking up his main proposition, Archimedes required two other geometric propositions, both proved by elementary geometry. They are the following:

1. The perimeter Q of an inscribed regular polygon is less than the perimeter C of the circle, and

2. the perimeter Q′ of the circumscribed regular polygon is larger than the perimeter C of the circle.

These two lead to the following inequalities of which Archimedes has made an extremely clever use in proving his proposition that

$$\text{Area (Circle)} = \tfrac{1}{2}r \cdot C$$

To get his first inequality he considered a regular inscribed polygon shown at the top of the next page.
Let AB be one of its sides and let its length be l. Let OM be a perpendicular on it from center O, so that AB is the base and OM the height of the triangle. If the length of OM be h, the area of triangle AOB is $(1/2)h \cdot l$. Since the polygon consists of n triangles each congruent with triangle AOB, the area of the polygon is

$$\frac{1}{2}h \cdot l + \frac{1}{2}h \cdot l + \cdots + \frac{1}{2}h \cdot l$$

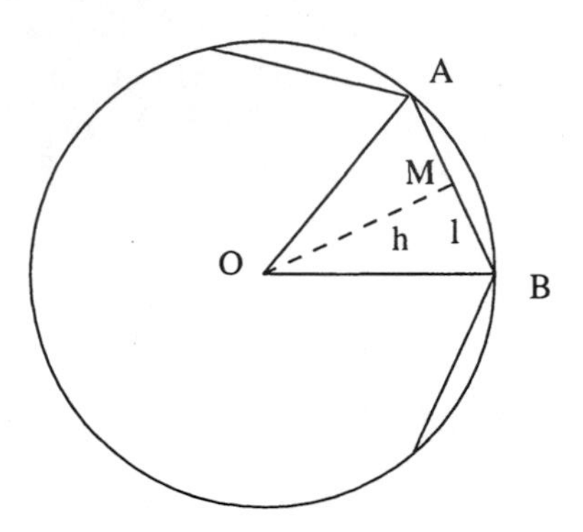

$$= \frac{1}{2}h \cdot (l + l + \cdots + l)$$

$$= \frac{1}{2}h \cdot Q$$

where Q is the perimeter of the polygon. Since Q is less than the perimeter C of the circle and h is less than r, the radius of the circle, he gets the first inequality

$$\text{Area(Inscribed Polygon)} = \frac{1}{2}h \cdot Q < \frac{1}{2}r \cdot C$$

Similarly, considering the circumscribing polygon with AB of length l as one of its equal sides, and OM $= r$ as the height of triangle AOB, he gets the area of the circumscribing polygon equal to (see figure on the next page.)

$$\frac{1}{2}r \cdot l + \frac{1}{2}r \cdot l + \cdots + \frac{1}{2}r \cdot l$$

$$= \frac{1}{2}r \cdot (l + l + \cdots + l)$$

$$= \frac{1}{2}r \cdot Q'$$

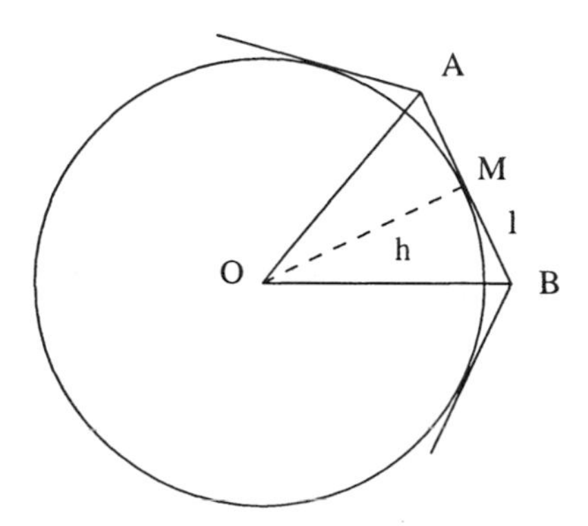

where Q' is the perimeter of the polygon. Since Q' is larger than the perimeter C of the circle, he gets his second inequality

$$\text{Area(Circumscribing Polygon)} = \frac{1}{2}r \cdot Q' > \frac{1}{2}r \cdot C$$

In his proof of the main theorem, given below, we write T for the product $(1/2)rC$. His above two inequalities now take the forms

$$\text{Area(Inscribed Polygon)} < T \tag{1}$$

and

$$\text{Area(circumscribing Polygon)} > T \tag{2}$$

The main proposition which Archimedes proved in his treatise *Measurement of the Circle* and its proof are as follows:

Proposition

> The area of any circle whose radius is r and whose perimeter is C is equal to the area of a right-angled triangle, whose one side about the right angle is r and the other is C.

Let the circle and the triangle be as in the figure on the next page. Let A and T be the areas of the circle and the triangle

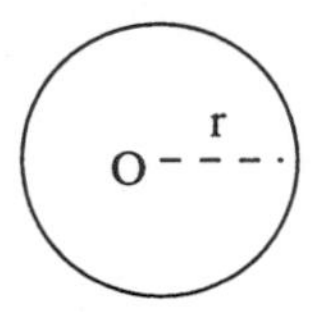

Radius = r
Perimeter = C

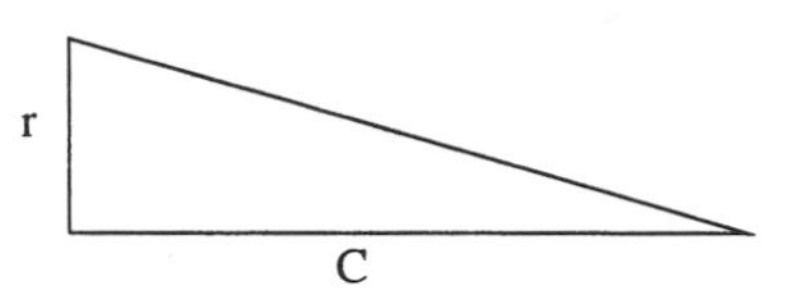

Sides about the right angle: r and C

respectively.
It is intended to prove that $A = T$.
If it is not so, then either $A > T$ or $A < T$.
First consider the case of larger A.

Case 1: $A > T$.

Since $A > T$, $A - T$ is a positive number. Therefore, there exists an inscribed regular polygon for which

$$A \text{ - Area(Inscribed Polygon)} < A - T$$

from which we get

$$T < \text{Area(Inscribed Polygon)}$$

But this contradicts inequality (1), namely

$$\text{Area(Inscribed Polygon)} < T$$

Therefore this case, $A > T$ is not tangible.
Now consider the case $A < T$.

Case 2: $A < T$.

Now since $A < T$, $T - A$ is a positive number; and a circumscribing polygon exists for which

$$\text{Area(Circumscribing Polygon)} - A < T - A$$

Adding A to both sides, this leads to

$$\text{Area(Circumscribing Polygon)} < T$$

But this contradicts inequality (2), namely

$$\text{Area(Circumscribing Polygon)} > T$$

Therefore even this case, $A < T$ is not tangible.
Since neither $A > T$ nor $A < T$ are possible, we conclude

$$A = T$$

as was to be shown.

The result established by Archimedes amounts to the formula

$$\text{Area of a circle} = \frac{1}{2}r \cdot C.$$

This formula is a land-mark amongst formulae for areas. But its real significance is not the actual formula but the method of getting the area of a given figure by *successive approximations which exhaust the error.* It was this method which ultimately led to *integral calculus* in the hands of later European mathematicians.

3. Quadrature of the Parabola.

Archimedes also gave another illustration of this same effective procedure by calculating the area of a parabolic segment.

Before we give details of this work we shall establish a property of a parabola which Archimedes has used in his proof. Archimedes has demonstrated the property by method of geometry current in his time. We shall give its demonstration by the use of methods of coordinate geometry.

Let the equation of the parabola shown on the next page be $y^2 = x$. Let AB be a chord with coordinates $(t_1^2,\ t_1)$ for A and $(t_2^2,\ t_2)$ for B.

Let C be a point on the parabolic arc AO, at which the tangent is parallel to AB.

Since $y^2 = x$, the slope of the tangent is $y' = 1/(2y)$ where y is the y-coordinate of C.

Since the tangent is parallel to AB, whose slope is $(t_1 - t_2)/(t_1^2 - t_2^2) = 1/(t_1 + t_2)$,

$$\frac{1}{2y} = \frac{1}{t_1 + t_2}$$

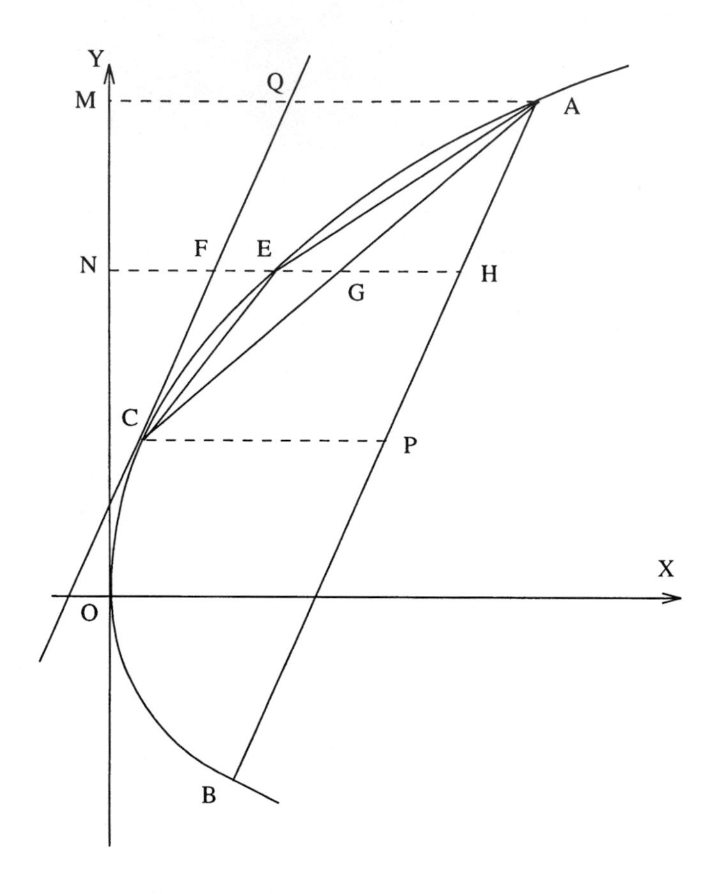

so that the y-coordinate of C is $\frac{1}{2}(t_1 + t_2)$, which is the same as the y-coordinate of the midpoint of AB. Therefore, if CP is drawn parallel to the x-axis, P would be the midpoint of AB.

Again, since the y-coordinate of C is $\frac{1}{2}(t_1 + t_2)$ and C is on the parabola $y^2 = x$, the coordinates of C are

$$\left(\frac{1}{4}(t_1 + t_2)^2,\ \frac{1}{2}(t_1 + t_2)\right)$$

The equation of the tangent at C, therefore, is

$$y - \frac{1}{2}(t_1 + t_2) = \frac{1}{t_1 + t_2}(x - \frac{1}{4}(t_1 + t_2)^2)$$

Let AM, the perpendicular from A on the y-axis meet this tangent at Q. Since the y-coordinate of Q is t_1, and since Q lies on the tangent, we get

$$t - \frac{1}{2}(t_1 + t_2) = \frac{1}{t_1 + t_2}(x - \frac{1}{4}(t_1 + t_2)^2)$$

Therefore $$x = \frac{1}{2}(t_1^2 - t_2^2) + \frac{1}{4}(t_1 + t_2)^2$$

Which, after reduction gives

$$x = \frac{1}{4}(3t_1^2 + 2t_1 t_2 - t_2^2)$$

And, since QA = AM − QM, we get

$$\text{QA} = t_1^2 - \frac{1}{4}(3t_1^2 + 2t_1 t_2 - t_2^2)$$

and, on reduction,

$$\text{QA} = \frac{1}{4}(t_1 - t_2)^2 \quad (3)$$

If E is a point on the parabolic arc AEC, and the tangent at E is parallel to chord AC, exactly as above, G would be the midpoint of chord AC, H would be the midpoint of AP and consequently would have its y-coordinate equal to

$$\frac{1}{2}[t_1 + \frac{1}{2}(t_1 + t_2)] = \frac{1}{4}(3t_1 + t_2)$$

This would also be the y-coordinate of E, whose x-coordinate, that is, EN would be

$$\frac{1}{16}(3t_1 + t_2)^2$$

Since the y-coordinate of F is $(1/4)(3t_1 + t_2)$, using the fact that it lies on the tangent at C, we get the x-coordinate of F, that is FN as equal to

$$\frac{t_1}{2}(t_1 + t_2)$$

From these two values, one of EN and the other of FN, we get the length FE which is EN − FN as

$$\frac{1}{16}(t_1 - t_2)^2 \qquad (4)$$

From equation (3) and (4), it follows that

$$\mathrm{FE} = \frac{1}{4}\mathrm{QA} = \frac{1}{4}\mathrm{FH}$$

Now AC is a diagonal and FH, a line joining the midpoints of the opposite sides of a parallelogram. They, therefore, bisect each other. Thus GH = (1/2)FH and since FE = (1/4)FH, we have EG = (1/2)GH. Therefore,

$$\begin{aligned}\text{Area (triangle AEG)} &= \frac{1}{2}\text{Area (triangle EGA)}\\ &= \frac{1}{8}\text{Area (triangle ACP).}\end{aligned}$$

Also,

$$\begin{aligned}\text{Area (triangle EGC)} &= \text{Area (triangle EGA)}\\ &= \frac{1}{8}\text{Area (triangle ACP).}\end{aligned}$$

Therefore,

$$\text{Area (triangle AEC)} = \tfrac{1}{4}\ \text{Area (triangle ACP).}$$

Similarly, if we choose point F on the parabolic arc BC, where the tangent is parallel to the chord BC, we shall get

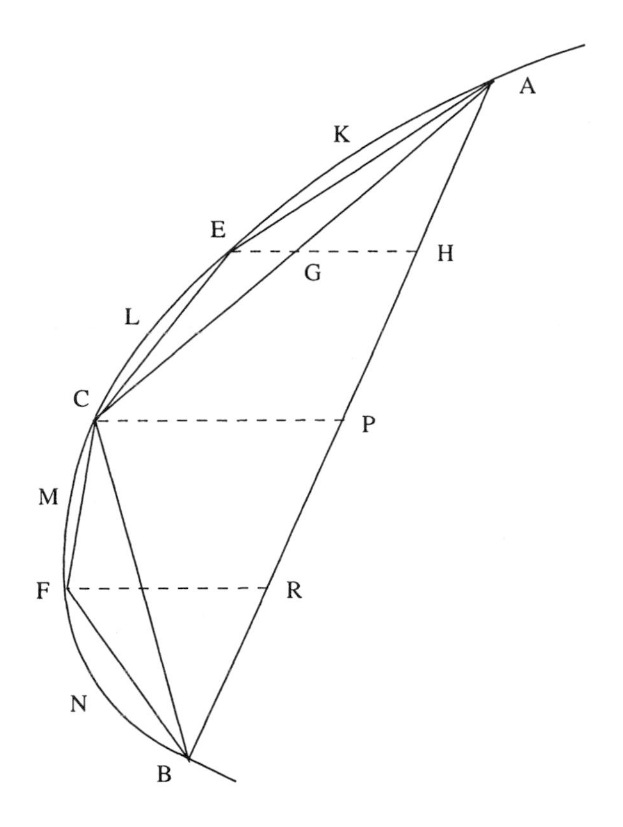

$$\text{Area (triangle BFC)} = \tfrac{1}{4}\text{Area (triangle BCP)}.$$

Adding these we get

$$\text{Area (triangle AEC)} + \text{Area (triangle BFC)}$$

$$= \tfrac{1}{4}\text{Area (triangle ABC)}.$$

This geometric result was just a means to the end which Archimed wanted to achieve. After establishing this result, he came to his real problem of finding the area of a parabolic segment, where during this latter process he once again illustrated the method of choosing such successive approximations as *exhaust* the error.

For this purpose, he chose the area of triangle ACB as his first approximation to the area of the parabolic segment ACB. In this approximation, (see the figure on the previous page)

$$\text{the error} = \text{area (AKELC)} + \text{area (CMFNB)}$$

He now chose area of figure AECFB as the next approximation to the area of the parabolic segment. Then

$$\begin{aligned}\text{the error} = \text{area (AKE)} &+ \text{area (ELC)} \\ &+ \text{area (CMF)} + \text{area (FNB)}\end{aligned}$$

which is *less* than the error in the earlier case. Therefore, by a successive choice of similar approximations obtained by repeating the application of the same technique, the *error gets exhausted.*

With this preliminary preparation, Archimedes turned to his main proposition about the area of a parabolic segment given below.

Proposition

If AB is the chord of a parabola, and C is a point on the parabolic arc where the tangent is parallel to the chord AB, then

$$\text{Area (parabolic segment ACB)} = \tfrac{4}{3}\ \text{Area (triangle ACB)}.$$

To prove this he used the same principle as he used in obtaining the expression $(1/2)rC$ for the area of a circle. In the earlier case he had constructed a chain of inscribed regular polygons as successive approximations to the area of the circle. In the present case, inscribed regular polygons were not relevant. Hence he devised another suitable type of polygon to inscribe in a parabolic segment. Triangle ABC was the first of them. (see figure on the next page). Choosing a point E on the parabolic segment at which the tangent is parallel to chord AC and F where the tangent is parallel to chord BC, he obtained polygon AECFB as the second. Choosing points G, H, K, L, on the parabolic arc at which tangents are parallel respectively to chords AE, EC, CF and FB, he obtained the polygon AGEHCKFLB as the next in the chain. And he continued to extend the chain so that these special polygons, step by step, cover more and more of the parabolic segment.

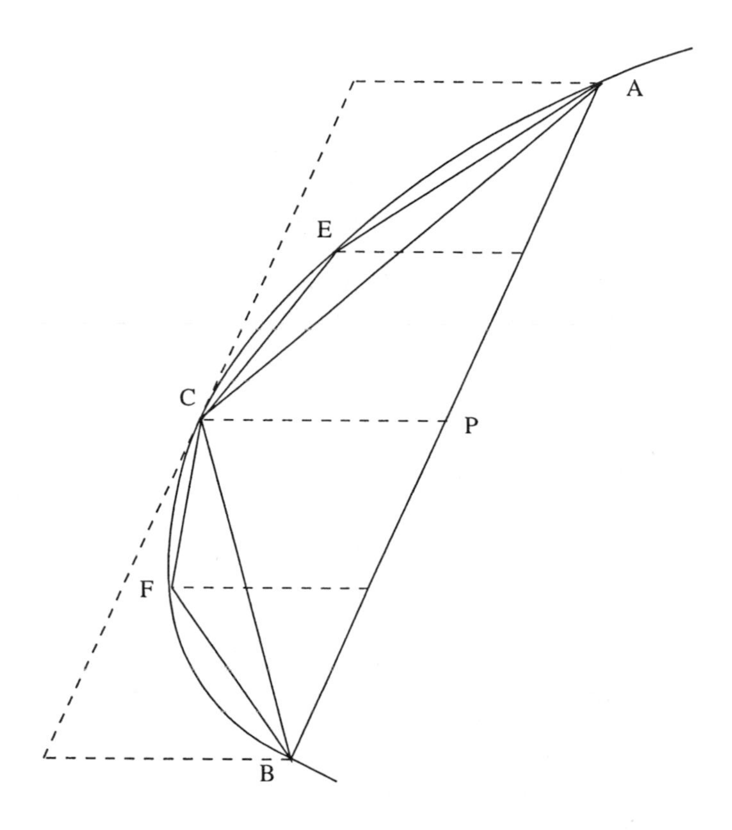

He used these special polygons as successive approximations to the area of the parabolic segment. He first showed that the *error* in these approximations can be *exhausted* in the sense which has been made clear above in the context of the measurement of a circle. He next showed that both the cases, namely,

$$\text{Area (Parabolic segment)} > \tfrac{4}{3}\ \text{Area (triangle ABC)}$$

and

$$\text{Area (Parabolic segment)} < \tfrac{4}{3}\ \text{Area (triangle ABC)}$$

lead to contradictions and hence concluded that

$$\text{Area (Parabolic segment)} = \tfrac{4}{3}\ \text{Area (triangle ABC)}$$

It is not necessary to give here all the details of the above steps. After all, these steps are either geometrical or algebraic manipulations. What concerns us here in the context of Integral Calculus is the methodology employed by Archimedes in arriving at the results through successive approximations, through showing that if proper figures are constructed for these approximations the errors involved could be exhausted, through showing how the visualized result can be obtained by a reasoning aptly called *double reductio ad absurdum.*

4. The work of Kepler.

Johannes Kepler came more than a thousand years after the last of the Greeks. He was born in 1571, almost four hundred years after the revival of mathematics in Europe after its sudden decline. He was really an astronomer and a student of the Danish astronomer Tycho Brahe. Tycho Brahe had an observatory of his own and had made and carefully recorded thousands of precise observations of the movements of the planets. When he died, this vast data passed on to the hands of Kepler. Kepler was extremely keen about his astronomical studies. When this enormous mass of observational data came into his hands, he took great pains to analyze the same. And from this analysis, he formulated what are since then known in the literature as *Kepler's Laws of Planetary Motion.*

The three laws are:

1. *The planets move around the sun in elliptical orbits with the sun at one focus.*

2. *The radius vector joining a planet to the sun sweeps over equal areas in equal intervals of time.*

3. *The square of the time of one complete revolution of a planet about its orbit is proportional to the cube of the orbit's semi-major axis.*

It is remarkable that one should be able to formulate such precise laws from empirical data. No other example of such astounding induction exists in science. Kepler was proud of this work of his and justly proud of it. In the preface to his book *Harmony of the Worlds* published in 1619 he writes

> I am writing a book for my contemporaries or – it does not matter – for posterity. It may be that my book will wait for a hundred years for a reader. Has not God waited for 6000 years for an observer?

What matters to us here in the present context of Integral Calculus is his second law.

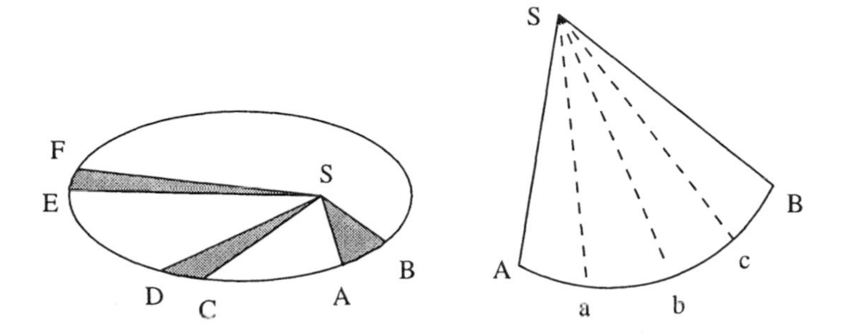

In order to conclude that areas SAB, SCD, SEF are equal he had to have some method of estimating these areas fairly satisfactorily. For this purpose he divided each of them into a number of sectors SAa, Sab, Sbc, ..., ScB, each so thin that their bases looked like bases of triangles. He calculated the areas of these extremely thin triangular slices by the usual formula for the area of a triangle.

This was probably the reason why Kepler was attracted to this subject of areas. He applied the same technique to the calculation of the area of a circle and even to the calculation of the volume of a sphere. In the case of the circle, as in the figure, he considered the arc-bases as straight lines, calculated the areas of various triangles by using the known formula for the area of a triangle and by adding up all the triangles, got the area of the circle as

$$\begin{aligned}\frac{1}{2}rb_1 \quad &+ \quad \frac{1}{2}rb_2 + \cdots + \frac{1}{2}rb_n \\ &= \quad \frac{1}{2}r(b_1 + b_2 + \cdots + b_n) \\ &= \quad \frac{1}{2}rC\end{aligned}$$

Similarly, to get the volume of a sphere, he divided the surface of the sphere into a number of small circular curved sections, considered these bases as flat circular bases, joined them all to the center of the sphere and got the sphere broken into "cones". He

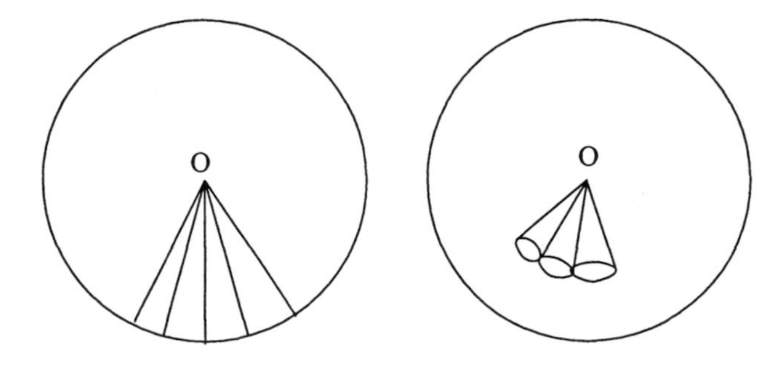

used the known formula for the volume of a cone and by adding up obtained the volume of the sphere as the sum

$$\begin{aligned}\frac{1}{3}ra_1 + \frac{1}{3}ra_2 + \cdots + \frac{1}{3}ra_n \\ = \frac{1}{3}r(a_1 + a_2 + \cdots + a_n) \\ = \frac{1}{2}r \cdot \text{surface area of the sphere}\end{aligned}$$

We find yet another illustration of his method in obtaining the volume of a wine barrel which he did as under.

He broke up the volume of the barrel into a number of cylindrical slices by planes parallel to the base. He considered them as

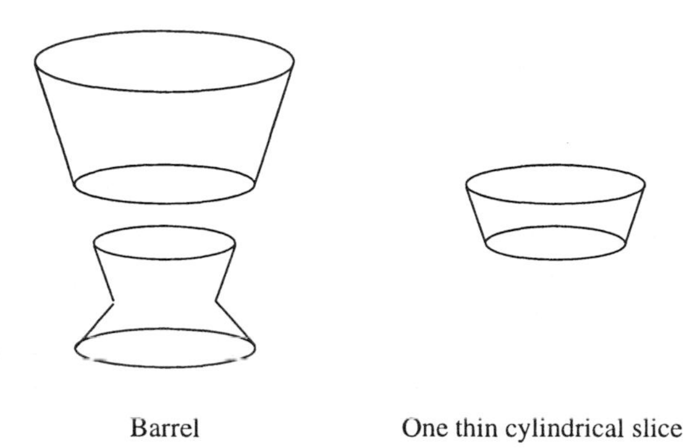

Barrel One thin cylindrical slice

perfect cylinders and using the known formula $\pi r^2 h$ for the volume of a cylinder he calculated the volume of each of these thin cylinders. He had to do it separately for each cylinder since the radius of their bases changed from one to the next because of the varying girth of the barrel from top to bottom. He took all the labor necessary and carefully measured all the girths C at various levels, calculated from C the value of r for each cylinder by using the known formula $r = C/2\pi$ and obtained the volume of each of the large number of the thin cylindrical slices. By adding them all together he obtained the volume of the barrel.

It must be noted that Kepler's methods of handling areas and volumes differed from that of Archimedes. Archimedes obtained the area of a circle or the parabolic segment by introducing a chain of suitable regular polygons successively approximating the area of the given figure. Kepler, on the other hand, broke up the given plane figure or solid into smaller parts of known forms such as triangles or cones or cylinder and obtained the area or volume required by adding up the areas or volumes of these smaller parts of known shapes. Kepler thus introduced a second major method of Integral Calculus.

5. The contribution of Cavalieri.

Cavalieri was an Italian, born in Milan in 1598. He was a student of Galileo and taught at the University of Bologna from 1629 until his death in 1647. The work for which he is now famous was first presented by him to the selection committee of the university which had met to consider him for a professorship at Bologna.

It was the custom then, as it is now in some universities, that prospective candidates for a teaching post are invited to give a seminar on the subject of their research. The famous 1920 talk of Professor G. H. Hardy given to the Oxford University on the eve of his occupying the Savillian Chair at Oxford was a talk of this type. And what is famous at *Erlanger Programme* was a talk by Professor Klein before he assumed the office of Professor of Mathematics at the University of Erlangen.

At the seminar which he gave to the selection committee of the University of Bologna, Cavalieri enunciated two basic principles, since then known as *Cavalieri's Principles*, and demonstrated their use in determining an area and a volume. Cavalieri's Principles are:

1. *If two planar pieces are included between a pair of parallel lines, and if the lengths of the two segments cut by them on any line parallel to the bounding lines are always in a given known ratio, then the areas of the two planar pieces are also in the same ratio.*

   and

2. *If two solids are included between a pair of parallel planes and if the areas of the sections cut by them on any plane parallel to the bounding planes are always in a given known ratio, then the volumes of the two solids are also in this same ratio.*

He demonstrated to the committee the extreme ease with which the area of an ellipse can be computed by comparing it with the area of a circle.

He did this in the following straight-forward way. We exhibit it here except that for easier and quicker understanding we do it by using a little of the standard analytic geometry.

Draw the ellipse

$$\frac{x^2}{a^2} + \frac{y^2}{b^2} = 1$$

and the circle

$$x^2 + y^2 = a^2$$

on the same set of coordinate axes with the two centers superimposed on each other and let them be *included* as Cavalieri's Principle wants it between the two vertical lines LM and UV. Now take

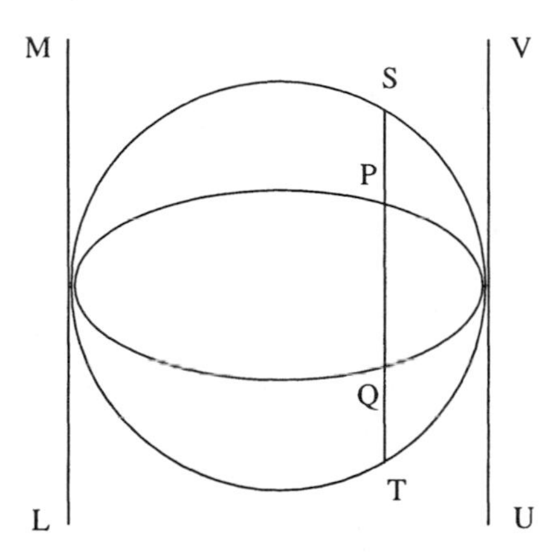

a section of the two curves by a line parallel to LM and UV. Since for the ellipse and the circle,

$$y = \pm\frac{b}{a}\sqrt{a^2 - x^2}$$

and

$$y = \pm\sqrt{a^2 - x^2}$$

respectively, the chord PQ of the ellipse has a length which bears a fixed proportion b/a to the length of the chord ST of the circle. By Cavalieri's principle it follows that

$$\text{area of the ellipse} = \tfrac{b}{a}\ \text{area of the circle.}$$

Cavalieri demonstrated his second principle by computing the volume of a sphere. For this purpose, along with the sphere of radius r, he took a cylinder of height $2r$ and radius r at base and top. He removed cones with vertices at the center of the cylinder

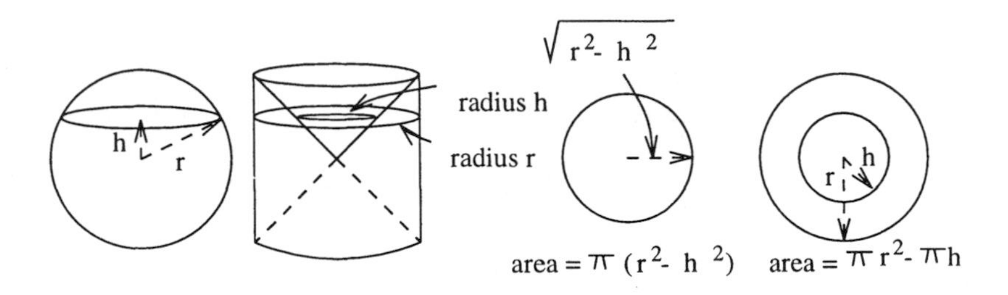

and the circular top and bottom of the cylinder as bases. He then placed the two solids – the sphere and the cones-deficient cylinders between two parallel bounding planes. When he took a section of these two solids by a plane parallel to the bounding planes and at distance h from their centers, he got a circle of radius $\sqrt{r^2 - h^2}$ as section of the sphere, and a ring with an outer circle of radius r and a hole of radius h in it, as shown in the two figures on the right. He showed that their areas were both equal. Hence, by the second principle, the volumes of the two solids would be equal. This means that

$$\begin{aligned}
&\text{the volume of the sphere}\\
&\qquad = \text{volume of the cone-deficient cylinders}\\
&\qquad = \pi r^2 \cdot 2r - \frac{2}{3}\pi r^2 \cdot r\\
&\qquad = \frac{4}{3}\pi r^3
\end{aligned}$$

It must have been an exciting experience for the selection committee. Cavalieri was selected and he stayed there for the next 18 years when his death removed him from the chair he so creditably adorned for 18 years.

In 1635, Cavalieri published all this work with applications to various areas and volumes in his book *Geometria Indivisibilium* (The Geometry of the Indivisibles). He explains his term *indivisible* in his book in the words

> A line is made up of points as a string is of beads; a plane area is made up of lines as a cloth is of threads; and a solid is made up of plane sections as a book is made up of pages.

His concept of a plane area as made up of lines and his basing the computation of a plane area on this concept amounts roughly to the formula

$$\int_a^b y\,dx$$

which we now use for computing plane areas. Cavalieri, however, has never shown any love for such general forms.

His book had had a great influence in England and France and must have made a lasting impression on the minds of Newton and Leibnitz.

6. The work of Fermat.

Pierre de Fermat was undoubtedly a great amature mathematician. Though his most valuable contribution to mathematics was in the field of Number-Theory, of which he was one of the founders, there is hardly any area of mathematics to which he has not made a contribution. He has his share in the creation of Analytic Geometry, Differential Calculus and Integral Calculus. In Differential Calculus he was the first to apply the method of Differential Calculus and find maxima and minima of algebraic functions. In Integral Calculus, he was the first to calculate the area under the curve $y = x^n$ by the method of erecting, as we do now do, thin rectangles approximating the area concerned. In fact, there is hardly any difference between what we now do and what he then did though he had at his time little support either algebraically or conceptually.

In order to obtain the area under the curve $y = x^n$, what he did was the following. He first drew the curve $y = x^n$ and then a chain

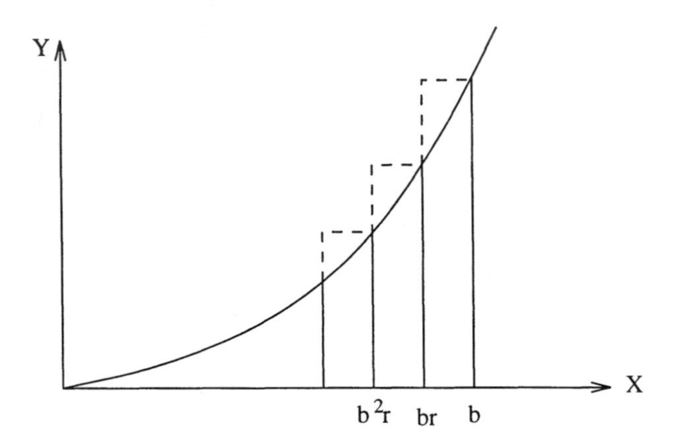

of outer rectangles. He introduced an infinite number of them and of decreasing widths, the widest being the one at the right end.

Let us denote by b, br, br^2, ... the abscissa of the base points with r a little less than 1. Thus, their widths, starting from the right are

$$b - br,\ br - br^2,\ br^2 - br^3,\ \ldots$$

that is

$$b(1-r),\ br(1-r),\ br^2(1-r),\ \ldots$$

Since the curve is $y = x^n$, the heights of these rectangles, from the right, are

$$b^n,\ b^n r^n,\ b^n r^{2n},\ b^n r^{3n},\ \ldots$$

The areas of these rectangles thus are

$$b^n b(1-r),\ b^n r^n br(1-r),\ b^n r^{2n} br^2(1-r),\ \ldots$$

and the total area of this chain of rectangles is

$$\begin{aligned}
b^{n+1}(1-r) + b^{n+1}r^{n+1}(1-r) + b^{n+1}r^{2(n+1)}(1-r) + \ \ldots \\
= b^{n+1}(1-r)[1 + r^{n+1} + r^{2(n+1)} + \cdots] \\
= \frac{b^{n+1}(1-r)}{1-r^{n+1}} \\
= \frac{b^{n+1}}{(1-r^{n+1})/(1-r)} \\
= \frac{b^{n+1}}{1+r+r^2+\cdots+r^n}
\end{aligned}$$

Letting r tend to 1, so that, the approximating rectangles become indefinitely thinner, we get this total area to be

$$\frac{b^{n+1}}{n+1}$$

As the approximating rectangles become thinner, the error involved in approximating the area under the curve gets *exhausted.* The final expression

$$\frac{b^{n+1}}{n+1}$$

is therefore the area that was being sought.

Fermat has used the same method as Archimedes did of introducing a chain of suitable figures to approximate the required area. He has however a chain which could suit any region. The final success would depend on whether or not the final summation involved is possible or easy or otherwise.

If n is not an integer but a rational number, a preliminary substitution $s = r^{1/q}$ reduces the summation to integral powers of s, and could be effected as above.

7. The work of Newton.

When Newton came to the problem of areas, it was already an old problem. A number of others had handled it: Archimedes, Kepler, Cavalieri, Fermat amongst others. Archimedes was the first who suggested that if the area of a standard closed figure, such as a circle or a parabolic segment is to be computed, the first step ought to be to construct a chain of suitable approximating polygons; the second step ought to be to show that the error committed in approximating the required area by the area of these polygons can be *exhausted*; and then finally do the appropriate calculations and get an expression for the area concerned.

All the others also have followed almost the same format but have made a small change here or a small change there, or have introduced in their work some procedural simplicity.

Newton was aware of all this work either through his reading or through the lectures of Isaac Barrow, his teacher at Cambridge. Barrow was almost the first to have noticed the genius in Newton. He gave to Newton everything he had, including even his Professorship at Cambridge – a gesture of magnanimity to be seen only in England.

Though aware of all the work of Archimedes and Kepler and Cavalieri and Fermat and even of some others, Newton did not follow the direction so well-set by these earlier masters. He was the first to have thought of calculating areas

by using an intrinsic property of areas.

The first example which he solved on areas was the following.

The area z from the origin to point x between a certain curve and the x-axis is given by

$$z = \left(\frac{n}{m+n}\right) a x^{\frac{m+n}{n}} \tag{5}$$

Find the curve.

He proposed this problem to himself and solved it somewhere around the year 1669. The tract in which he had included it was prepared in 1669 but published much later, around 1711.

The method he employed to solve it was the following. Let x be increased to $x + 0$ As the figure shows, the area now would be

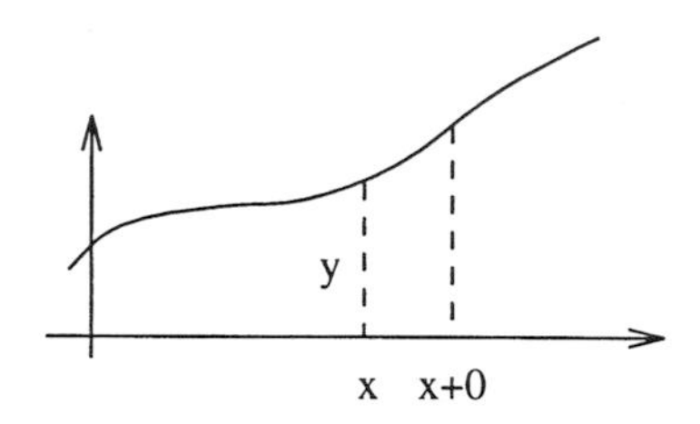

$z + 0y$ so that

$$z + 0y = \left(\frac{n}{m+n}\right) a(x+0)^{\frac{m+n}{n}} \tag{6}$$

Even before he took up this problem, Newton had discovered and established the *Binomial Theorem* for fractional indices. He now expanded the right side of (6) to get

$$z + 0y = \left(\frac{n}{m+n}\right) a \left[x^{\frac{m+n}{n}} + \frac{m+n}{n} x^{\frac{m+n}{n}-1} \cdot 0 \right.$$

$$\left. + \frac{1}{2!}\left(\frac{m+n}{n}\right)\left(\frac{m+n}{n} - 1\right) x^{\frac{m+n}{n}-2} \cdot 0^2 + \cdots \right]$$

Subtracting (5) from this, canceling 0 from the various terms and finally putting 0 equal to zero in all those terms which still contained 0 and powers of 0, Newton obtained the result that

$$y = ax^{\frac{m}{n}}$$

was the curve for which the area under the curve, above the x-axis and between the ordinates at 0 and x is

$$\left(\frac{n}{m+n}\right) ax^{\frac{m+n}{n}}$$

Newton applied this method to the quadrature of various other curves, such as

$$y = x^2 + x^{3/2} \quad \text{and} \quad y = \frac{a^2}{b+x}$$

By this solution, Newton established an important *new theorem.* By an example though, he showed that

Integration and Differentiation
are inverse processes.

Experience tells us that of the two processes, differentiation is the simpler one. Therefore all standard results of integration can be obtained by obtaining a related derivative.

Such were the beginnings of Integral Calculus. It was later developed much further and given a deeper conceptual setting by mathematicians like Cauchy, Riemann and Lebesgue.

This was Integral Calculus. In its development a number of mathematicians participated, including two all-time greats of mathematics, Archimedes and Newton. It started with one and ended with the other.

Exercises

1. $y^2 = x$ is a given parabola. Two points A and B on it have co-ordinates (49, 7) and (1, −1). Use the formula of Archimedes and show that the area of the parabolic segment cut off by the chord AB is 256/3.

2. AB is the chord of a parabola $x^2 = 4y$. The coordinates of A and B are (6, 9) and (−2, 1). Use the formula of Archimedes and show that the area of the parabolic segment cut off by AB is 64/3.

3. Use Fermat's method given in Section 5 of the chapter and calculate the area from $x = 0$ to $x = 6$ between the curve $y = x^5$ and the x-axis.

4. Use the method of Fermat to find the area from $x = 1$ to $x = 3$ between the curve $y = x^3$ and the x-axis.

5. Use Fermat's method to find the area from $x = 0$ to $x = 4$ between the curve $y = \sqrt{x}$ and the x-axis.

   [Hint: Take from right, 4, $4s^2$, $4s^4$, ... as points of division.]

6. Use Fermat's method to find the area from $x = 0$ to $x = 4$ between the curve $y = x\sqrt{x}$ and the x-axis.

   [Hint: Take the points to be 4, $4s^2$, $4s^4$, ...]

7. Apply Fermat's method to calculate the area from $x = 0$ to $x = 8$ between the curve $y = x^{1/3}$ and the x-axis.

8. (a) The area z from the origin to point x between a certain curve and the x-axis is given by

   $$z = \frac{2}{3}x^{3/2}.$$

   Find the area between $x = 4$ and $x = 9$.

   (b) Use Newton's method as in Section 7 and find the equation of the curve of part (a).

9. The area z from the origin to point x between a certain curve and the x-axis is given by

$$z = \frac{1}{4}x^4$$

Use Newton's method and find the curve for which this is true.

10. The area z from the origin to point x between a certain curve and the x-axis is given by

$$z = 3x^{\frac{4}{3}}$$

Use Newton's method and find the curve.

11. The area z from the origin to point x between a certain curve and the x-axis is given by

$$z = 4x^{\frac{8}{5}}$$

Find the curve.

Further Reading

1. Sir Thomas L. Heath, *A History of Greek Mathematics, Vol. II*, Clarendon Press, Oxford, 1921.

2. Carl B. Boyer, *The Concepts of the Calculus*, Hafner Publishing Company, 1949.

3. George F. Simmons, *Calculus Gems*, McGraw-Hill, Inc., New York, 1992.

CHAPTER FOUR

ANALYTIC GEOMETRY OF RENE DESCARTES

1. La Geometrie.

Rene Descartes was a talented French Mathematician of the seventeenth century. In 1637, he published a Discourse on the Method of Science and added to it an appendix of 116 pages, named LA GEOMETRIE. This appendix was translated from French into English in 1954 by D. E. Smith and Maria Leatham. In their preface to this translation, the translators pay their own tribute to Descartes as under:

> "If a mathematician was asked to name the great epoch-making works in his science, he might well hesitate in his decision concerning the product of the nineteenth century; he might even hesitate with respect to the eighteenth century; but as to the product of the sixteenth and seventeenth centuries, and particularly as to the works of the Greeks in classical times, he would probably have very definite views. He would certainly include the works of Euclid, Archimedes and Apollonius among the products of Greek civilization, while among those which contributed to the great renaissance of mathematics in the seventeenth century he would certainly include LA GEOMETRIE of Descartes and the PRINCIPIA of Newton."

La Geometrie is not a treatise by itself. It is an appendix to a much larger book of Descartes published in 1637 bearing the title which, translated in English, is "Discourse on the Method of Rightly Conducting Reason and Seeking Truth in the Sciences."

Though an appendix, it consists of 116 printed pages organized as three 'books' (as chapters were then called), entitled respectively "Problems the construction of which requires only Straight Lines and Circles", "On the nature of Curved Lines" and "On the construction of Solid and Supersolid Problems". His first book begins with the sentence

> "Any problem in geometry can easily be reduced to such terms that a knowledge of the lengths of certain straight lines is sufficient for its construction."

This is essentially the idea on which Descartes' *analytic geometry* is built. Descartes' geometry is not a new geometry. It is a *new method.*

2. Two illustrations.

We shall consider here two problems of geometry with the object of making clear the difference between Euclid's method and the method of Analytic Geometry advocated by Descartes.

> **Problem 1:** Construct a right-angled triangle whose hypotenuse is given and so is given the sum of its two legs.

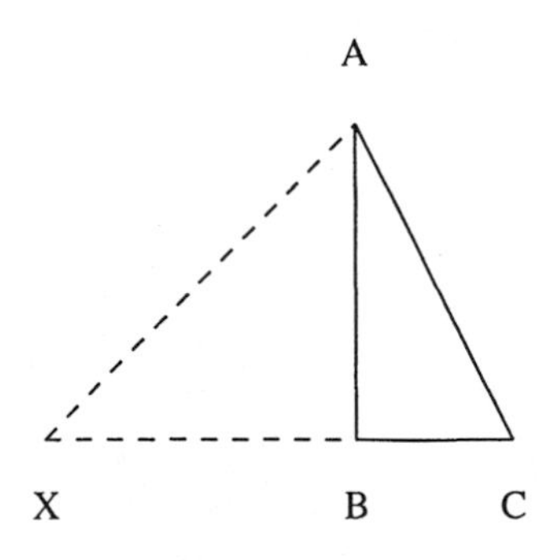

The Euclidean method of solving the problem would consist of a short analysis of the situation with a view to discover the kind of relation between the given parts and the parts required to construct the triangle. Let us *assume* that we have constructed the required triangle ABC. AC is given. Produce CB to a point X so that BX is equal to BA; CX equal to CB + BX is thus the second part given. Now it is easy to see that angle CXA is 45^0. Hence the following method could be suggested for the construction of the required triangle.

Draw a line XC equal to the given sum of legs. At X, draw

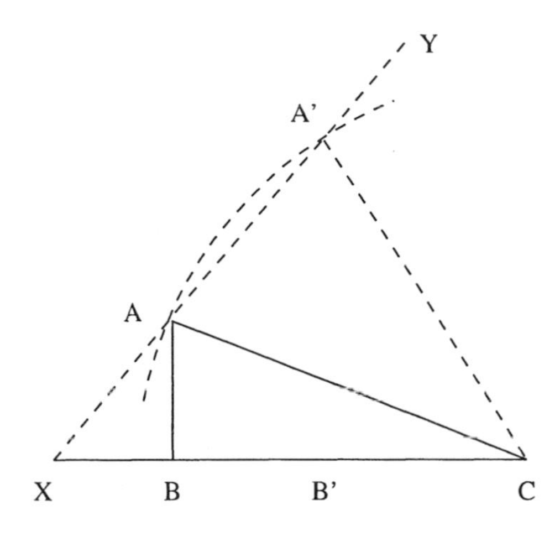

a straight line XY making angle CXY equal to 45^0. With C as center and the given length of the hypotenuse as radius, draw an arc of a circle to cut XY in the points A and A$'$. From A draw AB perpendicular to XC. ABC is the required triangle. Do the same with A$'$ and A$'$B$'$C would also be the required triangle with a different orientation but equal to the first in all respects.

The analytic method proceeds differently. To make it easy to follow, let us take it that the hypotenuse of the required triangle is 5 and that the sum of its two legs is 7. If the legs are taken to be x and y then one gets

$$x + y \qquad = 7$$

$$\text{and} \quad x^2 + y^2 = 25$$

so the given geometric problem, in the first instance, reduces to the algebraic one of finding x and y from these two equations and then constructing a triangle with three sides of lengths 5, x and y which two latter lengths are now known. Solving the above two equations, we get

$$x = 4, \quad y = 3 \quad \text{or} \quad x = 3 \text{ and } y = 4$$

corresponding to which two sets of values, we get two triangles ABC and A′B′C.

This is essentially Descartes' method of Analytic Geometry. Reduce the geometrical problem to one in Algebra. Solve this analytic problem by algebraic methods and present this algebraic solution back in the geometrical set-up of the original problem.

Consider another problem illustrating the same. This second problem brings out more clearly the simplicity of methods of Analytic Geometry.

> **Problem 2:** AB is a given straight line; C, a given point on it. Find a point D on AB produced such that
>
> $$\text{AD} \cdot \text{BD} = \text{CD}^2.$$

The analytic solution runs as follows: Let the figure below represent the situation. The given lengths are AC, CB. To find

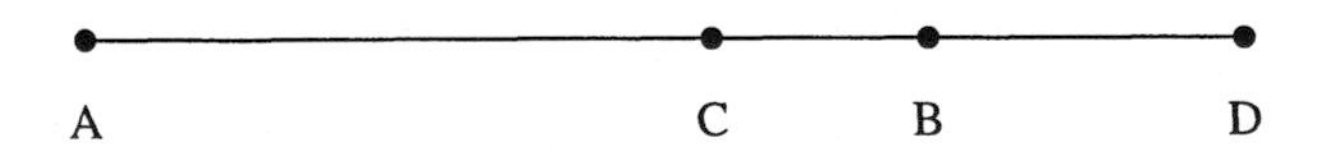

D is to find the length BD. Let

$$\text{AC} = a, \quad \text{CB} = b \text{ and the unknown length } \text{BD} = x.$$

Then

$$\begin{aligned} \text{AD} &= a + b + x \\ \text{BD} &= x \\ \text{and} \quad \text{CD} &= b + x \end{aligned}$$

so that AD·BD = CD^2 reduces to

$$(a + b + x) \cdot x = (b + x)^2$$

which is the algebraic form of the given geometric problem. From this last equation, we have to find x which is the required BD and with the knowledge of which we can find the point D on line AB. The last equation simplifies into

$$ax - bx = b^2$$

giving

$$x = b^2/(a - b).$$

A relevant question may be raised at this point. In the algebraic solution, we have freely used addition and multiplication and the answer turns out to be a division. Have these operations a meaning in geometry?

Descartes was quite aware of the possibility of such a question. The very second sentence in his first book is:

> Just as arithmetic consists of only four or five operations, namely, addition, subtraction, multiplication, division and the extraction of roots which may be considered a kind of division, so in geometry, to find required lines it is merely necessary to add or subtract other lines; or else taking one line which I shall call unity in order to relate it as closely as possible to numbers, and which can in general be chosen arbitrarily, and having given two other lines, to find a fourth line which shall be to one of the given lines as the other is to unity (which is the same as multiplication); or again to find a fourth line which is to one of the given lines as unity is to the other (which is equivalent to division); or, finally, to find one, two, or several mean proportionals between unity and some other line (which is the same as extracting the square-root, cube-root, etc., of the given line). And I shall not hesitate to introduce these arithmetical terms into geometry, for the sake of greater clearness.

3. Demonstration of arithmetic operations in geometry.

Up to and at the time Descartes was writing this, we were under a strong Greek legacy, under which, a product $AB \times CD$ meant the area of a rectangle, AB^2 meant the area of a square, AB^3 meant volume of a cube and higher powers such as AB^4, AB^5, ... were unintelligible as geometric forms. Such interpretations did not suit the new thinking of Descartes : for him, as he has briefly explained above, a^2 for example meant a line : a line obtained by constructing a third proportional to 1 and a; a^3 meant a line obtained by constructing a fourth proportional to 1, a and a^2; and so on for higher powers a^4, a^5, ... also. For him and his method they all had to be lines; and so also were products and square roots and cube roots etc. were also required to be lines.

Let us demonstrate how he proposed to multiply two lines so that the product also is a line. Let OX, OY as below be two inter-

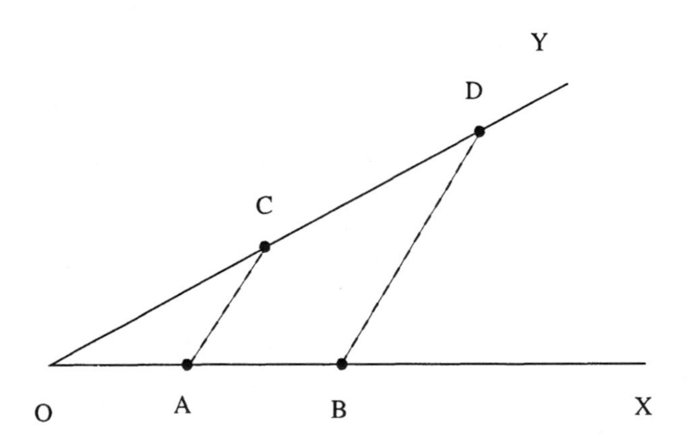

secting lines. Let OA be unity and let it be required to multiply OB by OC. Join AC and draw BD parallel to AC : then line OD is the *product* of OB and OC. (Because, since BD is parallel to AC, OD/OB = OC/OA and OA being unity, this gives OD = OB × OC.)

Similarly, if it be required to divide OG by OF and get the quotient as a line, join FG and from A, where OA is unity, draw

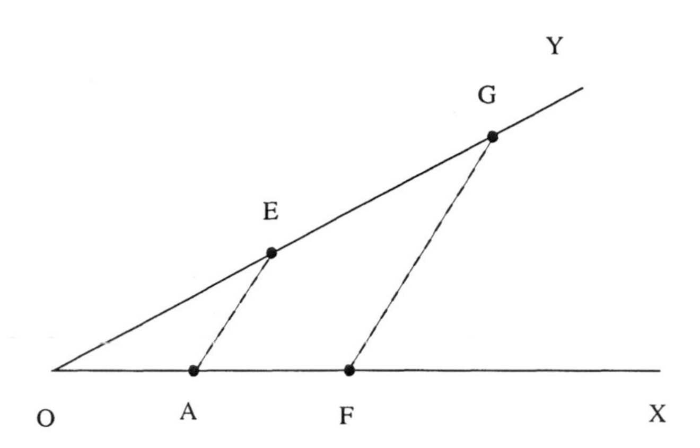

AE parallel to FG; then line OE is the *quotient* of OG by OF, (because OE/OA = OG/OF; giving OG/OF = OE.)

To obtain lines to represent powers a^2, a^3, a^4, ... one has merely to multiply a by a to get the line for a^2; multiply a by a^2 (for which the line has already been constructed) to get the line for a^3; multiply a by a^3 (the line for which has now been constructed) to get the line for a^4; and so on successively for all higher powers.

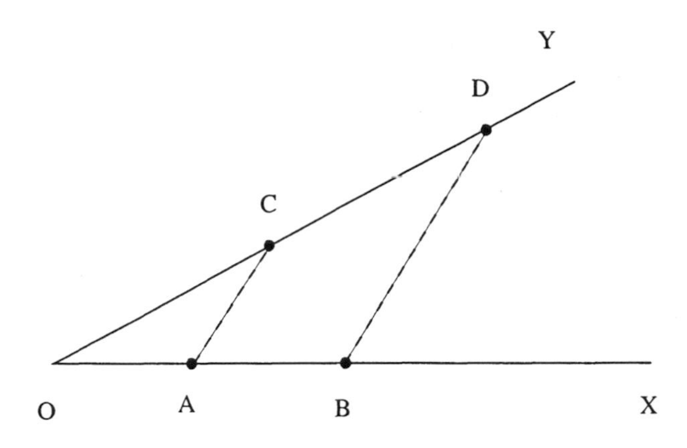

Let OA be unity, OC the line for a and OB, the line for a^n.

Then the line OD is the line for a^{n+1}.

If the line for the square-root of a given line LM is required draw LM and along the same line add MN equal to unity. Now

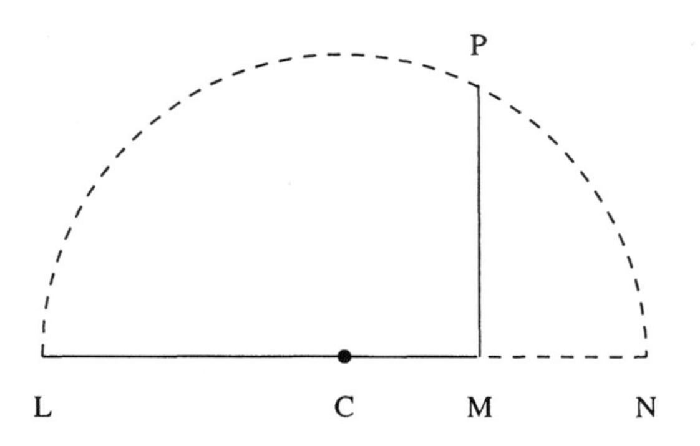

with the middle point C of LN as center draw a semi-circle. At M draw a perpendicular MP meeting the semi-circle at P. Then line MP is the line for the square-root of LM.

Thus Descartes showed that the five arithmetical operations of addition, subtraction, multiplication, division and extracting the square-root can be applied to the set of lines of geometry, and when so applied give lines as results. He further showed that all these constructions can be effected with the help of ruler and compass thus justifying the introduction of arithmetical terms in geometry.

4. The method of Analytic Geometry.

And it is at this stage that he returns to the theme of the general claim he had made in the very first sentence with which he begins his LA GEOMETRIE, namely:

> "Any problem in geometry can easily be reduced to such terms that a knowledge of the lengths of certain straight lines is sufficient for its construction."

He now elaborates the steps one must take to effect this reduction.

> If then we wish to solve any problem, we first suppose

> the *solution already effected*, and give names to *all lines* that seem needful for its construction – to those that are *unknown* as well as to those that are *known*. Then making *no distinction* between known and unknown lines, we must *unravel* the difficulty in any way that shows most naturally the *relations* between these lines, until we find it possible to express a single quantity in two ways. This will constitute an *equation*, since the terms of one of these two expressions are together equal to the terms of the other. We must find as many such equations as there are supposed to be unknown lines. ...
>
> If there are several such equations, we must use each in order, either considering it alone or comparing it with the others, so as to obtain a value for each of the unknown lines; and so we must continue them until there remains a single unknown line which is equal to some known line, or whose square, cube, fourth power, fifth power, sixth power etc., is equal to the sum or difference of two or more quantities.

Thus to solve a problem of geometry, Descartes recommends the following three clear steps:

ONE: Suppose that the solution is effected and give names to all the lines involved, both *known* and *unknown*;

TWO: Make no distinction between known and unknown lines and set up as many equations between them as there are unknown lines;

THREE: Eliminate the unknowns one by one and reduce the number of unknowns involved to the minimum.

It could be that this ultimately leads to *one* unknown, say x, and one equation which could be either linear such as

$$x = a$$

or a quadratic like

$$x^2 = -ax + b$$

or an equation in the unknown x of degree three or four or five etc. Then by solving the equation concerned, one can *determine* one or two or three ... possible values of the unknown x.

But it could happen that the system of equations set up as above does not finally reduce to one equation in one unknown but reduces to *one equation with more than one* unknown, such as

$$ax + by = c$$

or

$$ax^2 + bxy + cy^2 + dx + ey = f$$

etc. Then the solution is not determinate but consists of an infinite number of possible points satisfying this final equation and would then be a *curve*, a *locus*.

Consider the following problem.

> S is a given point and DR a given line not passing through S. P is a point in the plane of S and DR whose distance from S is equal to its distance from DR. Find P.

To use Descartes' method of analytic geometry, first consider the problem as solved and draw the figure showing in it all lines known and unknown. In the figure on the next page, S and DR are given; therefore SO, the perpendicular distance of S from DR is a known length. Let it be named a. SP is not known. Name it x. Now since P is such that its distance from S, namely SP, is equal to its distance from DR, namely NP, we get NP equal to SP. Therefore NP also is x. Now since the position of a point like P cannot be located by merely knowing that SP = NP, we *introduce* a line OX perpendicular to DR and passing through S. If MP is perpendicular to this determinate line, distance MP is unknown; name it y. Now OM being equal to NP and therefore to x and since OS = a, we get the following three equations:

$$\begin{aligned} \text{SP} &= x \\ \text{MP} &= y \\ \text{SM} &= x - a \end{aligned}$$

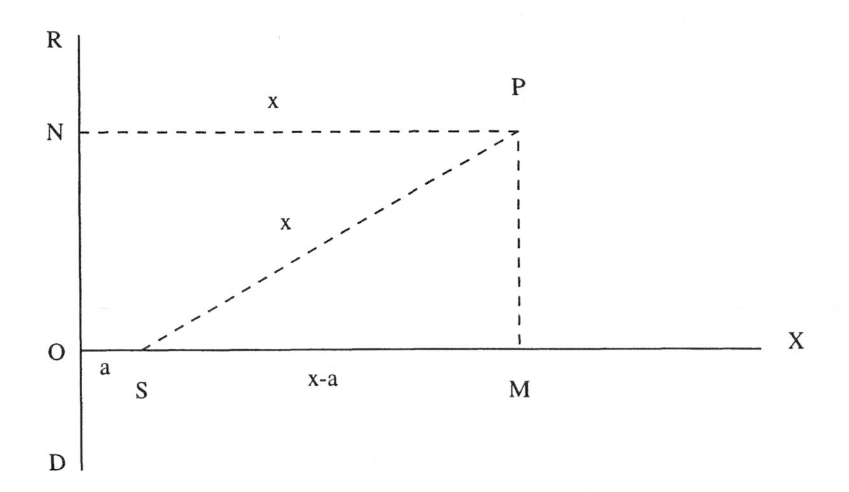

Triangle SMP is a right-angled triangle. Hence

$$y^2 + (x - a)^2 = x^2$$

which, when simplified, gives

$$y^2 = 2ax - a^2$$

This last equation cannot be further reduced, in the sense that neither x nor y can be removed from it and one equation involving one unknown cannot be obtained. This means that P is not a determinate point but is one of a set of an infinite number of points each one of which satisfies a kind of generic law given by the last equation. That is to say, in other words, that points P *lie on a curve*, which curve is the *locus* of a point which satisfies the given condition SP = NP.

Consider a second illustration.

> S and H are two given points. P is a point such that SP + HP is equal to a constant a. Find where P is situated.

Imagine that the solution is effected and that the solution is as is shown below. S and H are given; therefore distance SH is

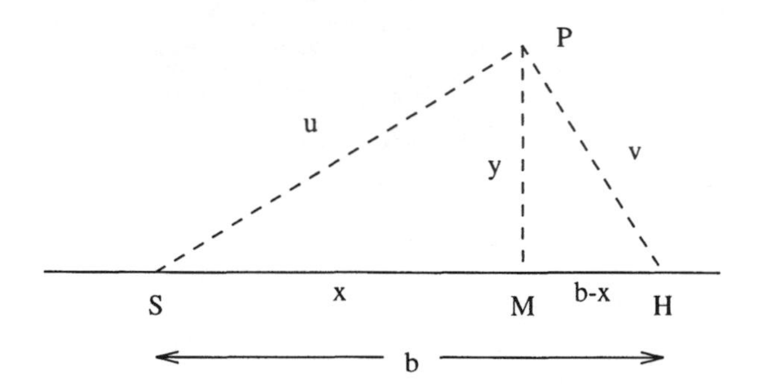

known. Name it b. SP and HP are unknown; name them u and v respectively. Since point P cannot be located merely from the given fact that SP + HP = a, we introduce line SH as a second line of reference. Let PM be a perpendicular on SH. This line MP is an unknown line; name it y. SM is an unknown line; name it x. Name line MH as z. We now have following equations connecting the known and the unknown lines.

$$u + v = a$$

$$x + z = b$$

$$\text{so that} \quad z = b - x$$

Now triangles SMP and PMH are right-angled; so that

$$u = \sqrt{x^2 + y^2}$$

$$v = \sqrt{(b - x)^2 + y^2}$$

These equations contain four unknowns, x, y, u and v. We first eliminate u and v and get the equation

$$\sqrt{x^2 + y^2} + \sqrt{(b - x)^2 + y^2} = a$$

which we now reduce with the help of algebra. Transferring the term $\sqrt{x^2 + y^2}$ to the other side and squaring both sides we get

$$(b - x)^2 + y^2 = a^2 + (x^2 + y^2) - 2a\sqrt{x^2 + y^2}$$

which simplifies into

$$b^2 - 2bx = a^2 - 2a\sqrt{x^2 + y^2}$$

transferring $-2a\sqrt{x^2 + y^2}$ to the left side, $b^2 - 2bx$ to the right side and squaring both sides, we get

$$4a^2(x^2 + y^2) = (a^2 - b^2)^2 + 4b^2x^2 + 4bx(a^2 - b^2)$$

which finally reduces, after further simplification, to

$$\frac{(x - a/2)^2}{(a^2 - b^2)/4} + \frac{y^2}{(a^2 - b^2)/4b^2} = 1$$

All this simplification was not necessary; it was done to see whether in the process, one of the two unknowns disappears. But it does not happen. This means that, as in the first illustration, P is not a determinate point but is one of an infinite number of them, each one of which satisfies generic law given in the last of the equations. P lies on a *curve* – on a *locus*; and the last equation could be said to be the *equation of this locus.*

5. The problem of Pappus.

The crowning glory of this method of *analytic geometry*, discovered and systematically exploited by Descartes is in his solution, by this method, of the famous unsolved Greek problem known as the problem of Pappus and which, as Pappus himself reports, could not be solved by Euclid, or Apollonius or the other reputed geometers of the Greek era.

The problem of Pappus is the following:

> $2n$ or $(2n + 1)$ lines are given in position. From a point C, lines are drawn to them at given angles. Find the locus of C if the product of n of these distances is equal to (or proportional to) the product of the remaining n (or $n + 1$) distances.

Descartes has illustrated his solution of the problem by considering four lines given in position. This is the first major problem to which Descartes has applied his new method of Analytic Geometry. And the way he has solved the problem is such that it can be easily extended if any number of given lines are involved in the problem. We shall give below Descartes' solution of the four-line problem elaborating and simplifying the same a little.

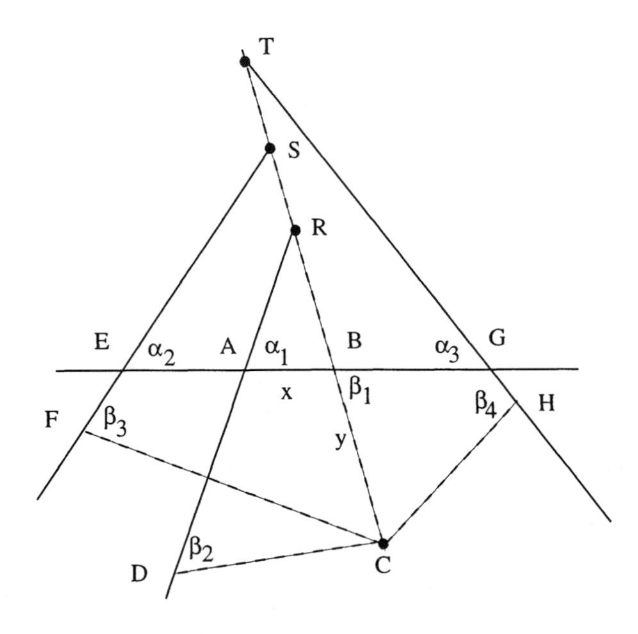

Here, EAG, RA, SE and TG are the lines given in position. According to his method, it is to be assumed that the solution is effected and C is one point belonging to the solution. CB, CD, CF and CH are drawn from C to the given lines at given angles. We find the locus of C if

$$\text{CB} \times \text{CF} = \text{CD} \times \text{CH}$$

Since lines EA, RA, SE and TG are given in position, distances EA, AG, SE, RA and TG are known: and for the same reason, the angles α_1 at A, α_2 at E, α_3 at G are known angles. Lines CB, CD, CF and CH are unknown; but since they are drawn at given angles, the angle β_1 at B, β_2 at D, β_3 at F and β_4 at H are known.

Distance AB is unknown; name it x. Distance CB is unknown; name it y.

We shall now calculate the unknowns, CB, CD, CF and CH one by one, in terms of the known and the two unknowns x and y.

Step 1: CB is already named y and so we have

$$\text{CB} = y \tag{1}$$

Step 2: Consider triangle ABR, two of whose angles are known; the third at R can be calculated; Let it be ν_1. Now apply sine-rule to the triangle ABR and get

$$\text{BR} / \text{AB} = \sin\alpha_1 / \sin\nu_1 = \lambda_1$$

so that $\quad \text{BR} = \lambda_1 \cdot \text{AB} = \lambda_1 x$

and $\quad \text{CR} = y + \lambda_1 x.$

Now in triangle CDR, angles at R and D are known. Therefore, applying sine-rule to it, we get

$$\text{CD}/\text{CR} = \sin\nu_1 / \sin\beta_2 = \lambda_2$$

$$\text{so that} \quad \text{CD} = \lambda_2 \cdot \text{CR} = \lambda_2(y + \lambda_1 x) = \lambda_2 y + \lambda_2\lambda_1 x \tag{2}$$

Step 3: Next consider triangle BSE. Length AE is known, say k; so that EB $= k + x$. Angles at E and B are known; therefore angle at S can be calculated. Name it ν_2. Now apply sine-rule to the triangle, so that, we get

$$\text{BS} / \text{BE} = \sin\alpha_2 / \sin\nu_2 = \lambda_3$$

so that $\quad \text{BS} = \lambda_3 \cdot \text{BE} = \lambda_3(k + x)$

and $\quad \text{CS} = y + \lambda_3(k + x)$

Now in triangle FCS, angles at F and S are known. Applying sine-rule to it, we get

$$\text{CF}/\text{CS} = \sin\nu_2 / \sin\beta_3 = \lambda_4$$

so that

$$\text{CF} = \lambda_4 \cdot \text{CS} = \lambda_4(y + \lambda_3 k + \lambda_3 x) = \lambda_4 y + \lambda_4\lambda_3 k + \lambda_4\lambda_3 x \tag{3}$$

Step 4: AG is known; call it l; so that BG $= l - x$.
Now in triangle BGT angles at B and G are known; therefore, the angle at T can be determined. Name it ν_3 and apply sine-rule to triangle BGT and get

$$\mathrm{BT}\ /\ \mathrm{BG} = \sin\alpha_3 / \sin\nu_3 = \lambda_5$$

so that
$$\mathrm{BT} = \lambda_5 \cdot \mathrm{BG} = \lambda_5(l - x)$$

and
$$\mathrm{CT} = y + \lambda_5 l - \lambda_5 x$$

Lastly, in triangle CHT, angles at T and H are known; so that applying sine-rule to the triangle, we get

$$\mathrm{CH}/\mathrm{CT} = \sin\nu_3 / \sin\beta_4 = \lambda_6$$

so that

$$\mathrm{CH} = \lambda_6 \cdot \mathrm{CT} = \lambda_6(y + \lambda_5 l - \lambda_5 x)$$

that is,
$$\mathrm{CH} = \lambda_6 y + \lambda_6\lambda_5 l - \lambda_6\lambda_5 x \tag{4}$$

Putting in a few abbreviations, we may write the values of CB, CD, CF and CH which are in the results above, numbered (1), (2), (3) and (4), as

$$\mathrm{CB} = y; \quad \mathrm{CD} = ay + bx;$$

$$\mathrm{CF} = c + dy + ex; \quad \text{and} \quad \mathrm{CH} = f + gy + hx$$

so that the condition

$$\mathrm{CB} \times \mathrm{CF} = \mathrm{CD} \times \mathrm{CH}$$

which these distances satisfy, becomes

$$y(c + dy + ex) = (ay + bx)(f + gy + hx)$$

which, when multiplied out and rearranged, takes the form

$$y^2 - (A - Bx)y - (Cx - Dx^2) = 0. \tag{5}$$

Solving this quadratic equation for y, we get

$$2y = (A - Bx) \pm \sqrt{Kx^2 + Lx + A^2} \tag{6}$$

Since the given condition on C gives in its final reduced form *one* equation and *two* unknowns, C describes a *curve*, a locus whose equation is given by (5) above.

Descartes has put the solution of this equation in the form of equation (6) and has given the descriptions of the loci which equation (6) represents as follows.

Case (i): If the expression under the radical sign is zero or a perfect square, the locus is a *straight line.*

Case (ii): If $K = 0$, the locus is a *parabola.*

Case (iii): If $K > 0$, the locus is a *hyperbola.*

Case (iv): If $K < 0$, the locus is an *ellipse.*

and if in equation (5), $B = 0$ and $D = 1$, the locus is a *circle.*

In fact, he has also shown how to work out the other details of the curves concerned.

Thus he completes the full solution of the problem of Pappus for four lines given in position. Two points may be noted about this solution.

ONE: He names CB and AB, the two unknown lines, as y and x; and using just the sine-rule with reference to two triangles, each time, he determines, one by one, the unknowns CD, CF and CH. Obviously, if there was a fifth line given and CI was drawn to it at a given angle, CI could have been calculated the same way as CD, CF and CH have been. And if a sixth line was there and CJ was drawn to it, its length also could have been similarly determined. And this would continue to hold good no matter how many lines were given in position initially.

TWO: The lengths of each of CB, CD, CF and CH are linear expressions in y and x. And since the same method is extended to the determination of CI, CJ, ..., their lengths also would be linear expressions in y and x. Therefore, the product of n of them would be an expression of degree n; and the product of $n+1$ of them would be an expression of degree $n+1$. And when such products are equated, as is the requirement of the problem of Pappus, the result would be *one* equation of degree n or $n+1$ in *two* unknowns and the *locus* of C would, in each case, be a *curve* of degree n or $n+1$.

Thus, the unsolved – almost mythical Greek problem of Pappus which had presumably defeated the efforts of even the greatest of the Greek geometers, who were using the sole method of straight edge and compass, could be solved by the new method of *analytic geometry.* The superior power of the new method was clear. This solution was, indeed, a great *tribute* to the method of Analytic Geometry and to Descartes, who was the first to discover it and systematically exploit it.

6. Standardized Analytic Geometry.

But this is not Analytic Geometry as we know it today. The term Analytic Geometry is currently employed in the context of standardized rectangular axes and distances x and y of the point under investigation measured from such rectangular axes. Under this scheme each *curve* which we know has a distinct *equation* corresponding to it; and reversely; every equation has a distinct curve corresponding to it. If the axes of reference are selected conveniently, the equations of the curves concerned turn out to be algebraically simple and elegant. The equations, more over, from their forms, frequently reflect such properties of the curves as symmetry. Thus, for example, if axes are suitably chosen, curves circle, ellipse, and parabola, get represented by equations

$$y^2 = a^2 - x^2$$

$$y^2 = (b^2/a^2)(a^2 - x^2)$$

and

$$y^2 = 4ax$$

One can notice even from the very forms of these equations, that the first two curves are *closed ovals*, symmetrical about a central point (closed ovals because $y^2 < a^2$ as well as $x^2 < a^2$, and symmetrical because x as well as y appear in even powers) and that the third curve is symmetrical about the x-axis but is *open* extending to infinity.

Also, from just a look at the first two equations, one can conclude that

the ordinate Y of the ellipse = (b/a) times the ordinate of the circle

and therefore can further conclude that

$$\int_0^a \mathrm{Y\ dx} = \frac{b}{a}\int_0^a \mathrm{y\ dx}$$

so that the area of the ellipse is (b/a) times the area of the circle, thus leading to the result

the area of an ellipse
of semi-axes a and b = (b/a) times the area of a circle of radius a
= $(b/a) \times \pi a^2$

Thus,

the area of an ellipse
of semi-axes a and b = πab

a result which could not have been established so elegantly otherwise.

One can indicate many other examples of the simplicity introduced in geometry by the method of analytic geometry. Consider, for example, the geometrical result

$$\mathrm{MP}^2 = \mathrm{AM} \times \mathrm{MB}$$

for a circle. To prove this in synthetic geometry, one method is to join PA and QB, prove triangles PAM and BQM similar, and from this similarly arrive at the conclusion

$$\frac{\mathrm{AM}}{\mathrm{QM}} = \frac{\mathrm{MP}}{\mathrm{MB}}$$

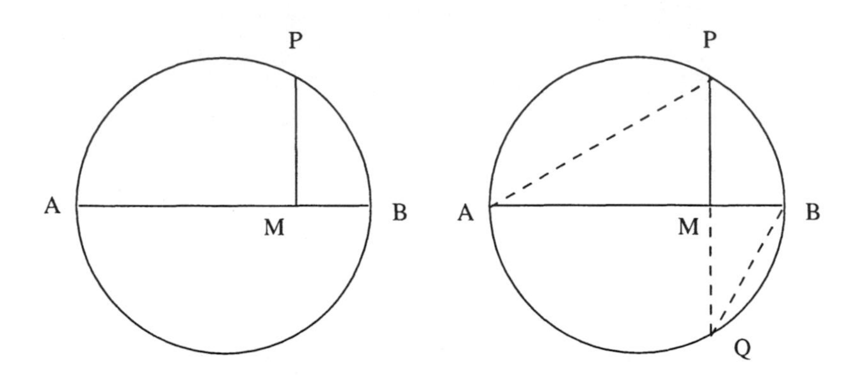

and further since diameter AB perpendicular to chord QP bisects it, we can express the preceding result as AM / MP = MP / MB and ultimately have it as

$$\mathrm{MP}^2 = \mathrm{AM} \times \mathrm{MB}.$$

Notice that for establishing this result, we introduced a construction, used two other geometrical properties of a circle, showed that two triangles are similar and then arrived at the final conclusion.

Now see how we arrive at the same conclusion, by the use of analytic geometry. Algebraically, a circle is

$$y^2 = a^2 - x^2$$

which on factorization of the right side gives

$$y^2 = (a + x)(a - x)$$

Now looking at the figure, one can see that if P is (x, y), then

$$y^2 = \mathrm{MP}^2, \quad a + x = \mathrm{AM} \quad \text{and} \quad a - x = \mathrm{MB}$$

so that, one has obtained the result

$$\mathrm{MP}^2 = \mathrm{AM} \times \mathrm{MB}$$

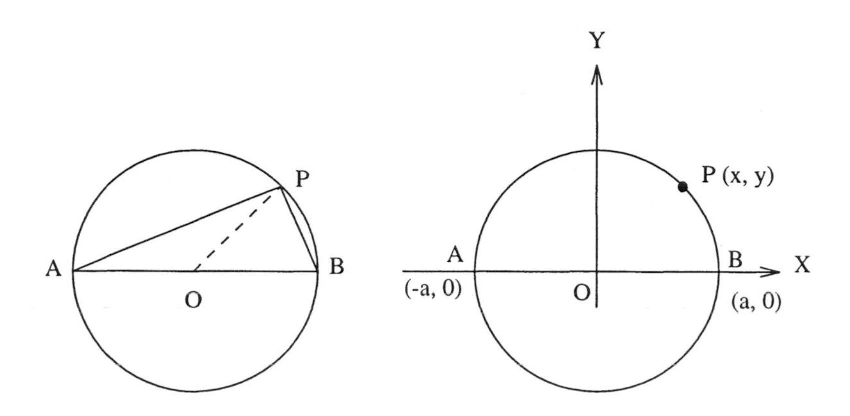

by employing such simple algebra.

Similarly, look at another illustration. We know that angle APB in a semi-circle is a right angle. One method to prove this result, in synthetic geometry, is to join PO, use results

$$\text{angle BOP} = \text{angle OAP} + \text{angle OPA} = \text{twice angle OPA}$$

$$\text{angle AOP} = \text{angle OBP} + \text{angle OPB} = \text{twice angle OPB}$$

and adding, get the required result

$$\text{angle APB} = \text{one right angle}$$

Notice how many steps we had to go through, and how many preliminary results we had to use; and now compare this whole preceding work with the following you have to go through for establishing the same result by methods of Analytic Geometry.

As we saw above, a circle is characterized by the equation

$$y^2 = a^2 - x^2$$

which on factorization and cross division, leads to the result

$$\frac{y}{x+a} \cdot \frac{y}{x-a} = -1$$

since the two ratios on the left are the *slopes* of PA and PB and since the product of these slopes is equal to -1, we conclude that

PA and PB are at right angles

just the result we were looking for, and arrived at by this method after so few and such simple algebraic steps and so little dependence on preliminary results.

Now, let us look at a situation of a different type – not of proving a result but of extending a result from two to three dimensions.

Consider a tetrahedron VABC in a three-dimensional rectangular frame of reference. Let the coordinates of V, A, B and C in this frame be respectively

$$(v_1,\ v_2,\ v_3),\quad (a_1,\ a_2,\ a_3),\quad (b_1,\ b_2,\ b_3),\ \text{and}\ (c_1,\ c_2,\ c_3).$$

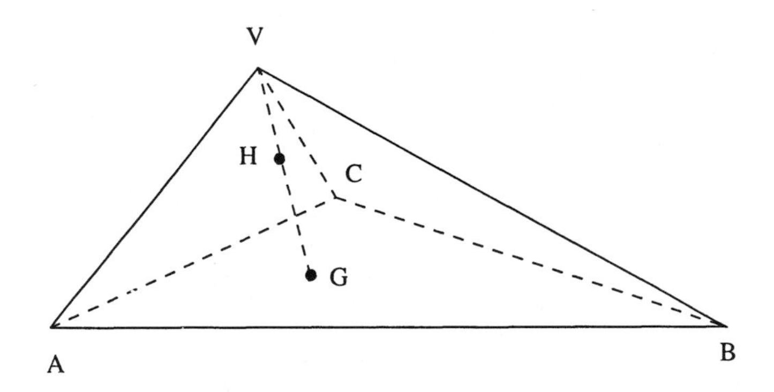

Let G be the centroid of triangle ABC; so that, as we know from plane geometry, its coordinates would be

$$\left(\frac{a_1+b_1+c_1}{3},\ \frac{a_2+b_2+c_2}{3},\ \frac{a_3+b_3+c_3}{3}\right)$$

Join VG and let H divide VG in the ratio 3:1. Then the coordinates of H would be

$$\left(\frac{a_1+b_1+c_1+v_1}{4},\ \frac{a_2+b_2+c_2+v_2}{4},\ \frac{a_3+b_3+c_3+v_3}{4}\right)$$

From the symmetry of this last result, we conclude that if every vertex was joined to the centroid of the opposite triangular face, all the four lines would meet in the same point H. Thus we have the theorem that

> The four medians of a tetrahedron meet in one point and that point divides each of them in the same ratio 3:1.

7. Analytic Geometry : the Method.

Standardized analytic geometry has thus a number of points of advantage over synthetic geometry. It is simple, and uniform and admits easy extensions. But Descartes never even thought of standardizing the method. And it is not surprising that Descartes paid no attention to this aspect. He was not writing for students: he had thought of an altogether novel way of looking at geometrical problems – of transforming them into algebraic forms – particularly those problems in the case of which even the best efforts of the great geometers of the Greek times had proved futile. His chief primary interest in the matter was to show how every geometry problem could be suitably transformed into the algebraic set-up.

Standardization of which we saw a few illustrations above came much later. When questioned on the point of the unknown distances which he names x and y and to which x and y or to only one of them he reduces all equations, he answered, in a letter to Princess Elizabeth:

> "In the solution of a geometrical problem, I take care, *as far as possible*, to use as lines of reference parallel lines or lines at right angles; and I use no theorems except those which assert that the sides of similar triangles are proportional, and that in the right triangle, the square of the hypotenuse is equal to the sum of the squares of the sides. I do not hesitate to introduce several quantities, so as to reduce the question to such terms that it shall depend only on these two theorems."

In his solution of the problem of Pappus, however, as shown

above, he used y for CB and x for AB; and CB and AB are not distances from perpendicular axes. In fact, he could have chosen A as origin, line AG as x-axis and a line through A at right angles to AG as y-axis. But these possibilities did not matter to him. These standardizations came later. Descartes never felt the need for them. His method was to choose some two unknown distances *conveniently*, name them x and y, express the other unknowns in terms of these two and the other known distances in the figure and finally utilize the given condition (such as CB $\times$ CF $=$ CD $\times$ CH above) to set up an *algebraic equation* satisfied by x, y and the other constants of the problem.

He would thus *transform* the problem of geometry into an *algebraic equation*, *solve* the algebraic problem algebraically and finally *invert* this solution back in geometry.

8. The Transform-Solve-Invert method.

The *transform-solve-invert* is a method quite frequently used in mathematics, often for proving and quite often for discovering results. Consider the following few illustrations.

Problem 1: Show that

$$x^7 - 7x^6 + 21x^5 - 35x^4 + 27x^3 - 9x^2 + 2x + 4 = 0 \tag{7}$$

has no root greater than 1.

To do this, put

$$x = y + 1 \tag{8}$$

in (7), which then becomes

$$y^7 + 2y^4 + 3y + 6 = 0 \tag{9}$$

and the given problem is *transformed* into the problem of showing that (9) has no root greater than 0, since as (8) shows, y has to be not greater than 0 if x is required to be not greater than 1. Now this latter problem is easy to show since if $y > 0$, the expression $y^7 + 2y^4 + 3y + 6$ would be greater than 0. Thus by *solving* this latter problem and *inverting* the conclusion, we arrive at the solution of the original problem.

Here, the device (8) of putting $x = y + 1$ has transformed the problem into a simpler problem – algebraic though it has remained – whose solution, when inverted, provided the answer of the original question.

Problem 2: Find a natural number x whose 23rd power leaves a remainder 3 when divided by 11.

In more sophisticated language, the problem is:

Find a natural number x such that

$$x^{23} \equiv 3 \pmod{11} \tag{10}$$

This is a problem about elements x of a multiplicative group modulo 11. We *transform* this problem to one about elements y of an additive group. The device we employ to get the appropriate group would be to set up an additive group isomorphic to the above multiplicative group. Under one such isomorphism, the corresponding elements run as below:

y :	0	1	2	3	4	5	6	7	8	9	(+, modulo 10)
$x = 2^y$:	1	2	4	8	5	10	9	7	3	6	(×, modulo 11)

Under this isomorphism, our original problem of solving (10) is *transformed* into the problem of finding y satisfying

$$23y \equiv 8 \quad (+, \text{modulo } 10) \tag{11}$$

Now, there are standard methods of solving linear congruence such as $23y = 8 + 10x$. In the present case, we solve (11) and get

$$y \equiv 6 \quad (+, \text{modulo } 10)$$

inverting the solution back into the set-up of the original problem, we get

$$x \equiv 9 \quad (\times, \text{modulo } 11)$$

as the solution required. This means that 9^{23} leaves a remainder 3 when divided by 11. Needless to add that all natural numbers and integers congruent to 9 modulo 11, such as 20, 31, 42, ..., −2, −13, ... also satisfy the equation.

Here we used the device of setting up an isomorphism for transforming the problem from the area of a multiplicative group to that of an additive group. Though even after the transformation, the area of the problem remained algebraic, it took us nearer the solution because it is easier to handle an additive group than a multiplicative group for situations as above.

This second illustration of the Transform-Solve-Invert method was surely more significant than the first. The third illustration below is, indeed, of quite some major significance. The situation considered in it, however, requires some mathematical preliminaries concerning the calculus of what is called the *Laplace transform* which we give here in short. We begin with a definition:

If $f(x)$ is a given function, the integral

$$\int_0^\infty e^{-px} f(x) dx$$

is called the Laplace transform of $f(x)$; it is written $L(f(x))$: it is a function of p.

Laplace Transform formulae for some functions:

1. $L(1) = \frac{1}{p}$

because, $L(1) = \int_0^\infty e^{-px}\, dx = \left[-\frac{e^{-px}}{p}\right]_0^\infty = 0 - (-\frac{1}{p}) = \frac{1}{p}$.

2. $L(x) = \frac{1}{p^2}$

because, $L(x) = \int_0^\infty e^{-px} x\, dx = \left[\frac{e^{-px}}{-p} x\right]_0^\infty - \int_0^\infty -\frac{e^{-px}}{p} \cdot 1\, dx$

$$= [0 - 0] + \frac{1}{p}\int_0^\infty e^{-px}\, dx$$

$$= 0 + \frac{1}{p}\, L(1) = \frac{1}{p^2}.$$

3. $L(\sin ax) = \frac{a}{p^2+a^2}$

because, $L(\sin ax) = \int_0^\infty e^{-px} \sin ax \; dx$

$$= \frac{1}{p^2+a^2}\left[-ae^{-px}\cos ax - pe^{-px}\sin ax\right]_0^\infty$$

using the known formula for such integrals. Putting $x = \infty$ and $x = 0$ one gets

$$L(\sin ax) = \frac{1}{p^2+a^2}[(-0+a)+(-0+0)]$$

$$= \frac{a}{p^2+a^2}$$

4. $L(\cos ax) = \frac{p}{p^2+a^2}$

This follows like (3) above.

5. $L(y') = -y(0) + pL(y)$

Let y be a function of x and let y', y'', ..., denote its derivatives with respect to x of orders 1, 2, ...

Then, $L(y') = \int_0^\infty e^{-px} y' \; dx$

$$= [e^{-px}y]_0^\infty - \int_0^\infty -pe^{-px}y \; dx$$

$$= [0 - y(0)] + p\int_0^\infty e^{-px}y \; dx$$

$$= -y(0) + pL(y)$$

6. $L(y'') = -y'(0) - py(0) + p^2L(y)$

Because, by 5. above, since y'' is the derivative of y',

$$L(y'') = -y'(0) + pL(y')$$

$$= -y'(0) + p(-y(0) + pL(y))$$

$$= -y'(0) - py(0) + p^2L(y)$$

7. And finally, since integration is a linear operator, we have

$$\mathrm{L}(\alpha f + \beta g) = \alpha\, \mathrm{L}(f) + \beta\, \mathrm{L}(g)$$

and if $f = g$, then $\mathrm{L}(f) = \mathrm{L}(g)$.

There is a lot more in the field of Laplace-Transforms: but what we have given above is enough for the point we are making in respect of the Transform-Solve-Invert method. Consider the following illustrations in this context.

Problem 3: Solve the differential equation

$$y'' + 4y = 4x \tag{12}$$

with initial conditions $y(0) = 1$ and $y'(0) = 5$.

Taking the Laplace-Transforms of both sides of equation (12) we get (12) *transformed* into

$$-y'(0) - py(0) + p^2\mathrm{L}(y) + 4\mathrm{L}(y) = \frac{4}{p^2}$$

and putting in it the given values of $y'(0)$ and $y(0)$, this becomes

$$-5 - p + p^2\mathrm{L}(y) + 4\mathrm{L}(y) = \frac{4}{p^2} \tag{13}$$

This is a simple algebraic equation in $\mathrm{L}(y)$: when *solved*, it gives

$$\mathrm{L}(y) = \frac{1}{p^2} + \frac{4}{p^2+4} + \frac{p}{p^2+4} \tag{14}$$

When *inverted*, this gives

$$y = x + 2\sin 2x + \cos 2x$$

the solution of the original differential equation.

This is, indeed, great – that a linear differential equation of the second order should be transformed by the method of the Laplace-Transform into a *simple linear algebraic* equation of degree one in $\mathrm{L}(y)$. But the greatness does not end here. In fact, *every* linear differential equation of *whatever order* it be, is *transformed* this way into a *simple linear algebraic* equation. Consider the linear differential equation

$$y''' + 4y' = 12x^2 + 6$$

with initial conditions: $y(0) = 4$, $y'(0) = 6$ and $y''(0) = -16$.

Let us drop details and only add here the two formulae

$$\mathrm{L}(y''') = -y''(0) - py'(0) - p^2y(0) + p^3\mathrm{L}(y)$$

and $$\mathrm{L}(x^2) = (2/p^3)$$

to the results established earlier.

When the given differential equation with the given initial conditions is transformed by the method of the Laplace-Transform, even in this case of a *third* order linear differential equation, we get the *simple linear algebraic* equation

$$p(4 + p^2)\mathrm{L}(y) = (24/p^3) + (6/p) + 6p + 4p^2$$

To solve this is easy: but quite some ingenuity is sometimes required to put the value of $\mathrm{L}(y)$ as a sum or difference of expressions in p which can be recognized as Laplace-transforms of known functions. Thus, in the case of the above equation we can put $\mathrm{L}(y)$ in the convenient form

$$\mathrm{L}(y) = \frac{6}{p^4} + 3\frac{2}{p^2 + 4} + 4\frac{p}{p^2 + 4}$$

so that, we get, when we *invert*,

$$y = x^3 + 3\sin 2x + 4\cos 2x$$

as the solution of the original differential equation.

This is surely a very impressive example of the use of the *Transform-Solve-Invert* method. But Descartes' method of *analytic geometry*, in which *every geometric problem*, is transformed into an *algebraic problem*, a problem from one area into a problem of an entirely distinct area is, indeed, more exciting. The difficulties in the former are algebraic and are met, if at all, at the stage where L(y) is to be expressed in a form convenient and suitable for writing down the inverses – a difficulty for the solution of which one needs to have at hand, a liberal supply of known Laplace-transform formulae. With Descartes' method, similarly, the difficulties could arise, if at all, at the algebraic stage. These difficulties could prove, occasionally, even insurmountable because of lack of sufficient knowledge of algebra. This is why Descartes has wisely devoted the entire Book Three of his LA GEOMETRIE to the subject of solution of algebraic equations, and has made his LA GEOMETRIE commendably complete and self-sufficient in as many respects as could be visualized in his time, the 1630s.

Descartes' work achieved three things: it

1. ultimately led to a simpler and uniform method of handling geometrical problems,

2. enabled the study of a large number of curves and surfaces, which were not amenable, till his time, to synthetic methods, and

3. provided to mathematics, in a major way, a *new method*, the *Transform-Solve-Invert* method.

As Howard Eves puts it:

> Far and away, the greatest, the most extensively developed, and the most productive instance of the transform-solve-invert technique yet devised by mathematics is that known as *analytic geometry*. There are few academic experiences that can be more thrilling to the student of elementary college mathematics than his introduction to this new and powerful method of attacking geometrical problems – for, analytic geometry is a *method*, rather than a branch of geometry.

Could there be a better appreciation of this epoch-making work of the great

Rene Descartes ?

Exercises

1. Lines of lengths a and b are given, $a > b$. Use the method indicated by Descartes and draw a line of length $b^2/(a - b)$.

2. LM, RS and UV are three parallel lines at distances 3 and 4 between them. C is a point not lying on these lines from where CB, CD and CH are drawn making angles 90, 45 and 30 degrees respectively with LM, RS and UV. Find the locus of C if CB $\times$ CH = CD^2.

3. In the figure in the text illustrating Descartes' solution of the problem of Pappus, let UV be a fifth line given in position, meeting CB produced in U beyond T and line EA in V beyond G. CI is drawn to it making a given angle β_0 with it. Find CI as a linear expression in x and y.

4. Use standardized analytic geometry and solve the following problem:
   AB, CD, EF and GH are four given lines whose equations are

   $$2y - 4x + 1 = 0, \quad y - x + 3 = 0,$$

   $$y - 2x - 2 = 0, \quad \textit{and} \quad y - 7x + 1 = 0$$

   respectively. P is a point from which PL, PM, PR and PS are drawn at right angles to AB, CD, EF and GH meeting them in L, M, R and S. Find the locus of P if

   $$\text{PL} \times \text{PR} = \text{PM} \times \text{PS}.$$

5. Show that in tetrahedron VABC, the centroid H bisects each of the three lines joining the midpoints of opposite edges VA, BC; VB, CA; and VC, AB.

6. Use the method of Problem 1 of Section 7 and show that

   $$x^5 - 10x^4 + 40x^3 - 80x^2 + 82x - 29 = 0$$

   has no root greater than 2.

7. (a) Show that the correspondence

0	1	2	3	4	5	6	7	8	9	(+, modulo 10)
1	7	5	2	3	10	4	6	9	8	(×, modulo 11)

   is an isomorphism between the multiplicative group modulo 11 and the additive group modulo 10.

   (b) Use this isomorphism and solve the congruence

$$x^{23} \equiv 3 \pmod{11}$$

   (c) Check whether you get the same answer as in the text even when a different isomorphism is used.

8. Use the method of Problem 2, Section 7, and find an integer whose forty-first power leaves a remainder 11 when divided by 13.

9. (a) Show that $\mathrm{L}(\mathrm{e}^x) = 1/(p-1)$

   (b) Solve the differential equation

$$y'' - y = 1, \quad y(0) = 0, \quad y'(0) = 1$$

10. (a) Show that

$$\mathrm{L}(y'''') = -y'''(0) - py''(0) - p^2y'(0) - p^3y(0) + p^4\mathrm{L}(y)$$

   (b) Show that

$$\text{(i)}\ \mathrm{L}(x^2) = \frac{2}{p^3} \qquad \text{(ii)}\ \mathrm{L}(\mathrm{e}^x) = \frac{1}{p-1}$$

   (c) Solve the differential system

$$y'''' = y - x^2$$

$$y(0) = -1, \quad y'(0) = 0, \quad y''(0) = 1, \quad y'''(0) = -2.$$

Further Reading

1. Rene Descartes, *La Geometrie*, translation by D. E. Smith and Maria L. Lathem, Dover, 1954.

2. Howard Eves, *Great Moments in Mathematics – Before 1650*, Mathematical Association of America, 1983.

3. C. R. Boyer, *History of Analytic Geometry*, Scripta Mathematica, 1956.

CHAPTER FIVE

THE METHOD OF DIFFERENTIAL CALCULUS

1. Apollonius : the concept of the minimum.

The method of Integral Calculus started with the Greeks: with Archimedes to be precise. He handled and solved the problem of finding certain areas, volumes and arc-lengths – problems which, in a natural way, occupy the minds of geometers. The method was essentially the method which we employ today in the process of integration: namely, that of finding an approximation to the measure concerned and then finding the limit to which this approximation formula leads. Though, neither the technical terms such as *limit* nor their precise definitions were then given, the reasoning employed was essentially the modern one.

The next advance on it came much later in the seventeenth century at the hands of Cavalieri and his method of the *indivisibles*. The story of this invention of Cavalieri, though a digression, is worth recalling here. Cavalieri had applied for the post of a Professor at the University of Bologna. That was in 1629. Bologna, then was a premier university of Europe. In those days, it was necessary for candidates for a Professor's post to give an oration before the interviewing committee about any piece of original work in which the candidate was then engaged. The method of *indivisibles*, for which Cavalieri earned a permanent place in the history of mathematics, was the subject of Cavalieri's oration for the post at Bologna. Cavalieri was selected.

Selection for a Professor's post was as rigid as that in those days in all universities and is still so in many western universities. Hardy's lecture to Oxford University in 1920 on the occasion of his selection in Oxford and Klein's famous Erlangen Programme of 1872 were orations of this kind.

Leaving these digressions apart, let us come back to the original point, that, Integral Calculus has a long history: it started with Archimedes of the third century B.C., followed by Cavalieri of the seventeenth century, followed thereafter almost immediately by Fermat, Wallis, Barrow, Newton and Leibnitz.

The method of Differential Calculus, relatively, had a much shorter history. Its use was made just in the seventeenth century, first, by Fermat. The use he made was in solving problems of maxima and minima and next in drawing tangents. Other contemporaries including Barrow, Newton and Leibnitz also limited their attention mainly to the same problems – problems of maxima, minima and tangents.

This suggests that they were very deeply influenced by their study of Greek geometry. It was the Greeks who introduced and developed the concept of a tangent to a curve. They did it for the circle first; and reaped a rich harvest. They were able to discover a number of important properties of a circle with the help of the tangent. Thoroughly impressed by these discoveries, Apollonius and others felt that the tangent and the normal supplied the main key to the discovery of properties of geometrical curves. Apollonius did for conic sections what Euclid had done for the circle. In his book on Conic Sections, there is a chapter on tangents quite early in the book: while his fifth chapter which deals with normals and curvature has the title "Maxima and Minima".

The procedure that Apollonius followed was quite interesting, though, *in no way was it a precursor* of the method of Differential Calculus. To draw the tangent was the basic objective of Apollonius: to do this, he calculated the *sub-normal* and the *sub-tangent*: and he used the concept of *minimum* to calculate the length of the sub-normal. The method of the minimum which he employed to calculate the length of the sub-normal is quite remarkable and is demonstrated below.

For this purpose, we have to recall that the maximum and minimum values of some simple functions can be obtained even without the use of calculus. Thus, since

$$1 - \cos(x) = 2\sin^2(x/2) \geq 0$$

the maximum value of $\cos x$ is 1. And since

$$1 + \cos x = 2\cos^2(x/2) \geq 0,$$

the minimum value of $\cos x$ is -1. And one can deduce, with a little more use of trignometry, the extreme values of some related functions also. Thus since

$$\sin x + \cos x = \sqrt{2}\cos(x - \pi/2),$$

the extreme values of $\sin x + \cos x$ are $\sqrt{2}$ and $-\sqrt{2}$.

Apollonius did not need these results: he used the following. Since, for any number x,

$$x^2 \geq 0,$$

the maximum value of $\lambda - x^2$ is λ and is attained when $x = 0$; and the minimum value of $x^2 + \lambda$ is λ and is attained when $x = 0$.

Now, since

$$px^2 - qx + r = p(x - \frac{q}{2p})^2 + r - \frac{q^2}{4p},$$

if p is positive, the *minimum* value of $px^2 - qx + r$ is *attained* when $x = q/(2p)$.

This is the result which Apollonius employed in determining the length of the *sub-normals* at points on the Conic Sections. His method is demonstrated below for a point on the Parabola, using unlike Apollonius, the language of modern analytic geometry.

Let P(x, y) be the point on the parabola $y^2 = 4ax$ at which we want thc normal; (or else, let G be the point on the axis *from which* we want the normal to the parabola). Apollonius defined the normal GP to be the *shortest distance* from G to the curve: probably, he must have argued to himself that being perpendicular to the tangent line at P, GP is the shortest distance from G to the tangent line and hence to the curve.

Now,

$$\begin{aligned}
\mathrm{GP}^2 &= y^2 + (\alpha - x)^2 \\
&= 4ax + \alpha^2 - 2\alpha x + x^2 \\
&= x^2 - 2(\alpha - 2a)x + \alpha^2 \\
&= [x - (\alpha - 2a)]^2 + \alpha^2 - (\alpha - 2a)^2
\end{aligned}$$

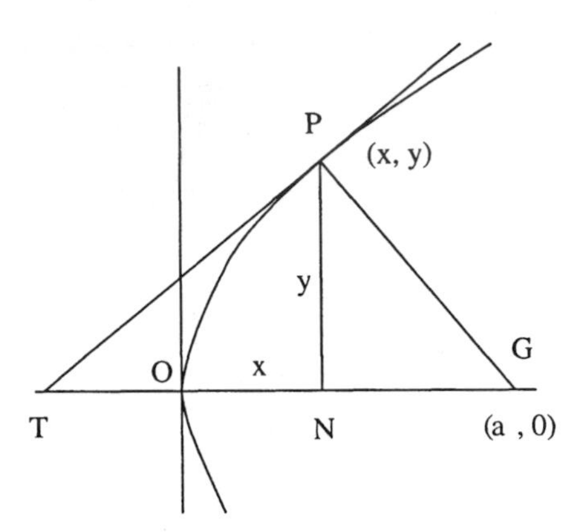

Hence GP is a *minimum* when $x - (\alpha - 2a) = 0$. That is, when α - $x = 2a$, that is, when NG $= 2a$.

To draw the Normal at P, take G on the axis such that NG $= 2a$; and join GP : that would be the normal at P.

The sub-tangent $t =$ TN can be obtained from the similarity of triangles NTP and NPG which gives us

$$\frac{t}{y} = \frac{y}{2a}$$

leading to

$$t = y^2/(2a) = (4ax)/(2a) = 2x.$$

Apollonius determined the lengths of the Sub-normals and Sub-tangents at points on the ellipse and the hyperbola and drew the Normals and Tangents at these points.

The method was indeed interesting; but it depended on the method to find the *minimum* value of a quadratic expression. For doing this, mere algebra was enough; but the same would not have been possible if the expression involved was a polynomial of higher degree.

This certainly was not differential calculus. But these ideas, these ways of looking at tangents and normals, the fact that the method entirely depended on the ability to find the *minimum* and the fact that tangents play an important role in the discovery of properties of curves, very greatly influenced the route which discoverers of the method of differential calculus took later in the seventeenth century.

2. Roberval and Descartes.

Locating the shortest distance from a point to a curve, that is, finding the minimum value of a certain length was the guiding idea behind Apollonius's method. Apollonius could do this because he was handling curves which, in our present language we call curves of the second degree. But when in 1637 Descartes published his essay on Analytical Geometry, every algebraic equation, of whatever degree it be, became a *curve* and all curves became *algebraic equations* of varied degrees. And Apollonius's method of finding the minimum ceased to be applicable. Therefore, it became necessary to find either a new method for determining the length of a sub-normal, or an entirely new method for drawing tangents and normals. A method applicable to all functions and to all curves was necessary and workers in the field were all keenly in search of such methods.

There were two ways in which Greek work influenced the later mathematicians. Some argued to themselves that since the ultimate aim was to draw a tangent to a curve, why not concentrate all thinking and all effort only on this problem? Why worry about sub-normals, sub-tangents, maxima or minima or other similar considerations? May be, they worked for the conics: what is the guarantee that they will succeed in other cases, in cases of curves of higher degrees?

This group consisted of such workers as Roberval and Descartes and they did achieve success but not to any appreciable degree. We shall briefly go over an example each from the works of Roberval and Descartes.

Roberval conceived that a curve is generated by a point whose motion is compounded of two known motions. Then the resultant of the two velocity vectors of the two conceived motions gives the tangent line to the curve. For example, in the case of a parabola, we may consider the two motions as away from the focus and away

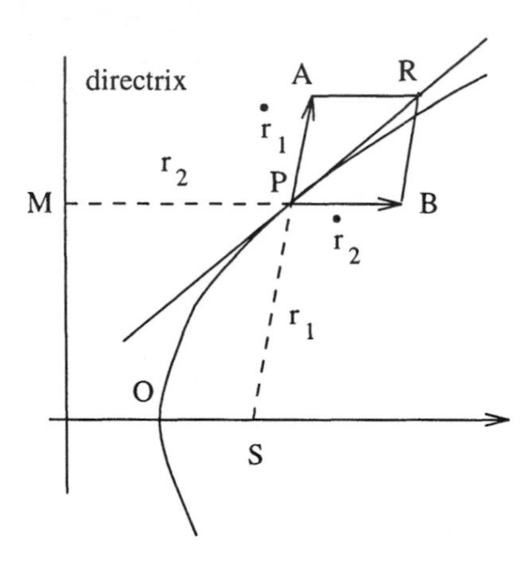

from the directrix. Since the distances r_1 and r_2 of the moving point from the focus and the directrix are always equal in a parabola, the velocity vectors $\dot{r}_1$ and $\dot{r}_2$ of the two motions would be equal in magnitude. The resultant vector would, therefore, bisect angle APB. Since the resultant vector is to be along the tangent line, it follows that in a parabola, the bisector of the angle between the focal radius SP and the perpendicular MP from the directrix is the tangent line at P.

The method appears attractive but is obviously of limited application. To conceive in the case of each given curve the two appropriate motions is not easy. The method, anyway, is not the method of differential calculus and hence need not be pursued here any further.

Descartes did not concentrate on the tangent; he suggested a method for finding a normal. Suppose the curve is $f(x, y) = 0$ and suppose that $P(x_1, y_1)$ is the point on the curve where we want the normal. Let $Q(x_2, 0)$ be a point on the x-axis. Then the equation of the circle with Q as centre and QP as radius, is

$$(x - x_2)^2 + y^2 = (x_1 - x_2)^2 + {y_1}^2. \tag{1}$$

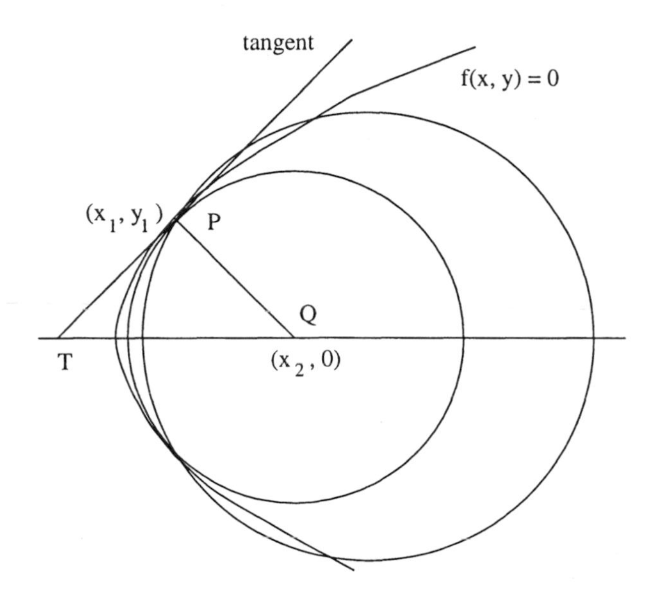

To find the x-coordinates of the points in which this circle intersects the given curve $f(x, y) = 0$, we eliminate y between the equation $f(x, y) = 0$ and the equation (1) of the circle. The elimination leads to an equation in x. Now determine x_2 such that this last equation in x has a pair of roots equal to x_1. The circle will now be tangent to the curve $f(x, y) = 0$ at the point (x_1, y_1), and QP will be the normal to $f(x, y) = 0$ at P.

We shall illustrate Descartes' method by finding the normal at (-1, 2) on the ellipse $3x^2 + y^2 = 7$. The circle with $(x_2, 0)$ as center and QP as radius is

$$(x - x_2)^2 + y^2 = (x_2 + 1)^2 + 4.$$

Where it meets the ellipse $3x^2 + y^2 = 7$, we have

$$(x - x_2)^2 + (7 - 3x^2) = (x_2 + 1)^2 + 4,$$

which simplifies into

$$x^2 + x_2 x + (x_2 - 1) = 0.$$

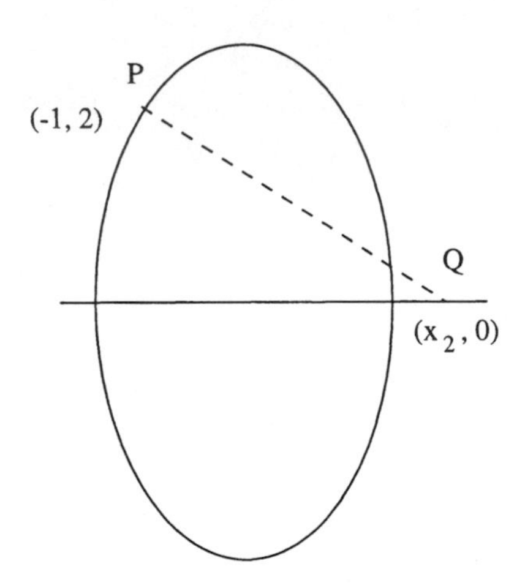

If this has equal roots,

$$x_2{}^2 = 4(x_2 - 1), \quad \text{giving} \quad x_2 = 2.$$

Thus the line joining (2,0) to (−1,2) would be the normal at (-1,2).

3. Fermat : the method of differential calculus.

The work of Roberval and Descartes could, in no way, lead to the method of differential calculus. Just as Roberval and Descartes devoted themselves wholly to the problem of drawing a tangent to a curve, some others clearly discerned that the king-pin of the whole problem was to discover a *general method* to attain the same objective; either through a sub-normal or a sub-tangent or in a way different from both. It is these *seekers* of a *method* who discovered the *method of the differential calculus.*

And the first amongst these discoverers of the Method of Differential Calculus was Fermat. And the first explicit demonstration of the method by Fermat was in problems of maxima and minima.

He must have been thinking about these problems for quite some time; but it took a definite shape when he saw a casual remark by Kepler. Kepler was an astronomer and for years engaged in analyzing the vast observational data collected and classified by his senior, the astronomer Tycho Brahe. While analyzing the data about the apparent movements of planets on a celestial sphere, he found that when a planet arrives at an extreme position while turning from direct to retrograde or retrograde to direct motion, the coordinates of the planet *remain unchanged* for a small period of time. And Kepler made a remark that the increment of a function becomes vanishingly small when at an extreme position. When Fermat saw this remark, it must have struck him as the possible key to unlock the secret of maxima and minima. He turned Kepler's observation upside down and made it a provisional criterion for an extreme value. He must have tried it on a few functions and must have felt convinced about it.

And he stated the criterion as under:

If $f(x)$ has a maximum or a minimum at x then, when e is small, the value of $f(x+e)$ is almost equal to the value of $f(x)$. Therefore he advocated the following procedure to find the maximum of a function:

1. Put tentatively $f(x+e) = f(x)$.
2. Make the equality correct later by putting $e = 0$.
3. The roots of the resulting equation give the values of x for which $f(x)$ is a maximum or a minimum.

The first problem in which Fermat demonstrated his procedure was the problem of dividing a number Y into two parts x and $Y-x$ such that the product $x(Y-x)$ of the two parts has a maximum value.

To solve this, he wrote $f(x)$ for $x(Y-x)$ and took the first step in his procedure by setting $f(x+e) = f(x)$, that is by putting

$$(x+e)(Y-x-e) = x(Y-x),$$

which on simplification becomes

$$-xe + e(Y-x) - e^2 = 0.$$

As his second step he now divides by e, and gets

$$-x + (Y-x) - e = 0.$$

In this, he finally puts $e = 0$ and gets

$$-x + (Y - x) = 0, \quad \text{i.e.,} \quad x = Y/2.$$

so that, to get the product of the two parts maximum, each part should be $Y/2$, and the maximum product would be $Y^2/4$.

What has to be done is:

- $f(x+e)$ has been put equal to $f(x)$.
- The difference $f(x+e) - f(x)$ is divided by e.
- In this new expression e is put equal to 0.

From the resulting equations the value of x at which $f(x)$ attains a maximum or a minimum value is obtained.

Fermat's logic may not be strictly satisfactory; and the inherent shortcomings, that his procedure gives only a necessary condition and is not definitive on whether the value obtained is a maximum or a minimum, is surely unerstandable. But the *novel* approach to the problem that he advocated is indeed *remarkable.* It gave the world of mathematics a *new method,* a powerful method which gave a radical turn to the development of mathematics, the *method of differential calculus.*

Let us look at the procedure once again from a slightly more advanced point of view. The procedure starts with

$$f(x+e) = f(x).$$

If we now use Taylor's theorem, this leads to

$$f(x) + ef'(x) + \frac{e^2}{2!}f''(x) + \ldots = f(x).$$

So that when simplified, one gets

$$ef'(x) + \frac{e^2}{2!}f''(x) + \ldots = 0$$

and we find that e is a factor of the left side whatever be the function f we start with. Therefore Fermat's instruction to "divide by e" is realisable in every case. After dividing by e, one gets

$$f'(x) + \frac{e}{2!}f''(x) + \ldots = 0$$

in which, except for the first term, every other term has e as a factor; so that putting $e = 0$ in it, the equation reduces to the *correct* equation

$$f'(x) = 0$$

for determining extreme points.

Fermat thus stands vindicated; not in respect of logic ; but surely in the matter of the procedure he advocates for finding the extreme values of a function.

Fermat also devised a general procedure for finding a tangent line to a curve. He did it by showing a way to get the sub-tangent at the point concerned unlike Apollonius who did it with the help of the sub-normal and the use of maxima or minima. The first step he took was the following. Let $f(x, y) = 0$ be the curve, P(x, y) be

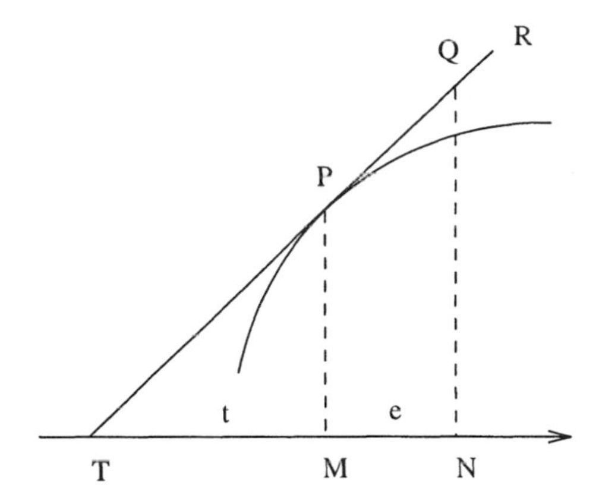

a point on it and TPR the tangent line at P. Let the sub-tangent TM be t. Let Q be a point on this tangent line, near point P. Let Q be (X, Y) and MN $= e$, an infinitesimal which ultimately is to be put equal to 0. From similarity of triangles QTN and PTM, one gets QN/PM = TN/TM, giving $Y/y = (t+e)/t = 1 + e/t$. So that $Y = y(1 + e/t)$, and the coordinates of Q are $(x + e,\ y(1 + e/t))$.

The procedure for finding t which Fermat advocates at this stage is:

First put $f(x+e,\ y(1+e/t)) = f(x,\ y)$ and then simplify, divide by e, put $e = 0$ and solve the resulting equation for t. With this value of t, locate point T, join TP and this would be the tangent line at P.

We illustrate the procedure by applying it to draw the tangent to the ellipse $2x^2 + 3y^2 = 35$ at the point (-2, 3) on it. As a first

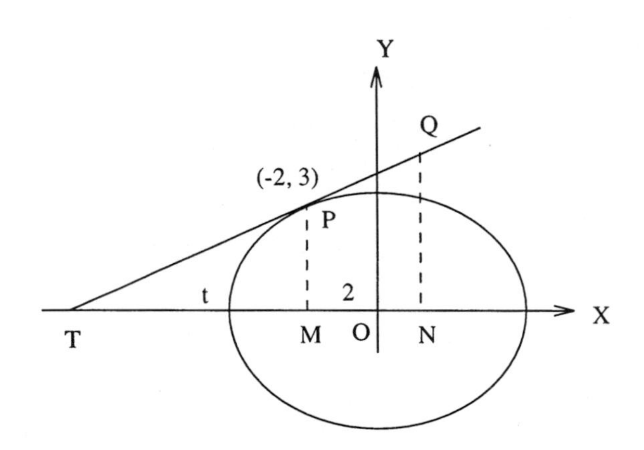

step, put

$$2(-2+e)^2 + 3 \cdot 3^2(1+e/t)^2 = 35.$$

This simplifies into

$$-8e + 2e^2 + 54e/t + 27e^2/t^2 = 0.$$

Division by e gives

$$-8 + 2e + 54/t + 27e/t^2 = 0.$$

Putting $e = 0$ leads to the equation

$$-8 + 54/t = 0,$$

finally giving $t = 27/4$.
Therefore, the x-coordinate of T is $-(27/4+2) = -35/4$.

Locating T with this knowledge, join TP to get the tangent line TP at P.

Let us examine the method as we did earlier. Using Taylor's theorem on the left of Fermat's starting equation

$$f(x+e,\ y+ye/t) = f(x,\ y),$$

we get

$$f(x,\ y) + e\frac{\partial f}{\partial x} + \frac{ey}{t}\frac{\partial f}{\partial y} + \frac{1}{2!}\left[e^2\frac{\partial^2 f}{\partial x^2} + 2c\frac{ey}{t}\frac{\partial^2 f}{\partial x \partial y} + \frac{e^2y^2}{t^2}\frac{\partial^2 f}{\partial y^2}\right]$$

$$+ \ldots = f(x,\ y),$$

which simplifies into

$$e\frac{\partial f}{\partial x} + e\frac{y}{t}\frac{\partial f}{\partial y} + \frac{e^2}{2!}\left[\frac{\partial^2 f}{\partial x^2} + 2\frac{y}{t}\frac{\partial^2 f}{\partial x \partial y} + \frac{y^2}{t^2}\frac{\partial^2 f}{\partial y^2}\right]$$

$$+\frac{e^3}{3!}[\ldots] + \ldots = 0.$$

Since e is a common factor, division by e is realisable, and when done leads to

$$\frac{\partial f}{\partial x} + \frac{y}{t}\frac{\partial f}{\partial y} + \frac{e}{2!}[\ldots] + \frac{e^2}{3!}[\ldots] + \ldots = 0$$

If we put $e = 0$ in this, we get the equation

$$\frac{\partial f}{\partial x} + \frac{y}{t}\frac{\partial f}{\partial y} = 0$$

leading to the value

$$t = -\frac{\partial f/\partial y}{\partial f/\partial x}y,$$

a perfectly *correct* result, once again establishing the validity of the procedure advocated by Fermat to get the value of the sub-tangent to a curve; a second significant use of the method of Differential Calculus.

4. Professor Isaac Barrow and his method.

The second mathematician to contribute substantially to the Method of Differential Calculus was a professional, Professor Isaac Barrow, Professor at Cambridge and a teacher of Newton. He has given the method in his 1669 book "Lectiones Opticae et Geometricae". Though he also gave a method to draw a tangent-line to a curve and actually applied it to a number of curves, the procedure he advocated was more directed to finding the equal of what we now call the *derivative* of a function.

Let us illustrate.

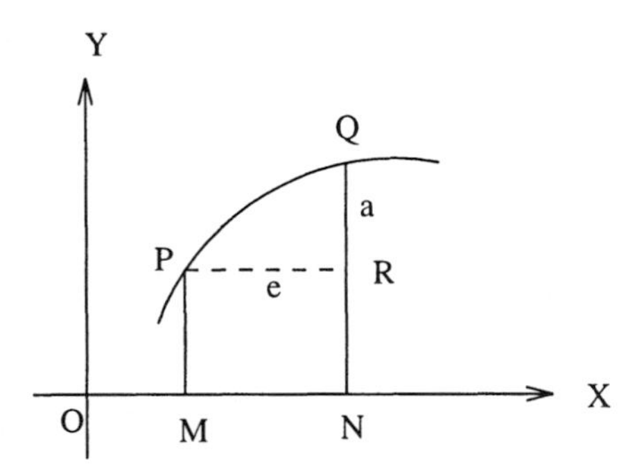

Let P be a point on

$$f(x,\ y) = 0 \tag{2}$$

with coordinates $(x,\ y)$. Let a point Q be chosen on the curve near P and let its coordinates be $(x + e,\ y + a)$, where e and a are both infinitesimally small. Since Q is on the curve, we would have

$$f(x + e,\ y + a) = 0. \tag{3}$$

Barrow, at this stage, suggests the following steps:

- Simplify (3) with use of (2).
- Neglect squares and higher powers of e and a.
- Solve the resulting equation for a/e.

This *change of emphasis* from finding the value of the subtangent to the determining of the value of a/e was *new* and a *major* contribution of Barrow to the streamlining of the Method of Differential Calculus. As we will see below, the value of a/e obtained from $f(x+e, y+a) = 0$ by neglecting squares and higher powers of e and a is the *derivative* of the function at x.

Let us apply Barrow's procedure at point (x, y) on

$$x^3 + y^3 = n\,xy \tag{4}$$

and determine the value of a/e at the point. Here, as per Barrow, the first step is to write

$$(x+e)^3 + (y+a)^3 = n(x+e)(y+a). \tag{5}$$

Opening out, and using (4) to simplify it, we get

$$3x^2e + 3xe^2 + e^3 + 3y^2a + 3ya^2 + a^3 = nxa + nye + nea.$$

On neglecting the terms in e^2, a^2 and ea gives us

$$3x^2e + 3y^2a = nxa + nye.$$

Finally, solving for a/e, we get

$$\frac{a}{e} = -\frac{3x^2 - ny}{3y^2 - nx}.$$

One may verify that this, indeed, is the value of the derivative of function y given by (4).

Barrow's real contribution is in drawing attention to the number a/e. Yet, he also, like others, puts his method to use in drawing tangents to given curves.

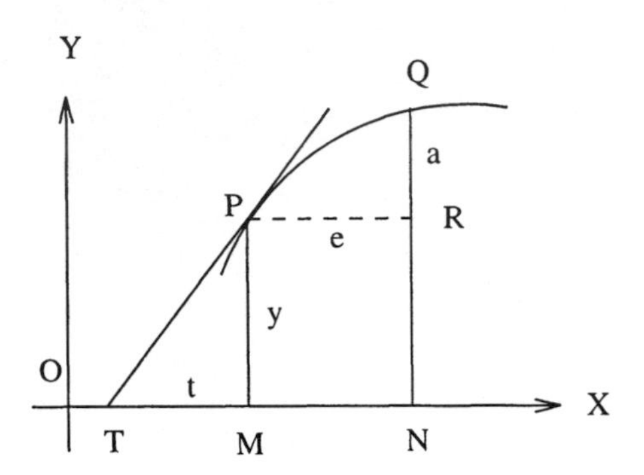

Thus, let P and Q be neighboring points on the given curve. Consider, tentatively, the figure QPR as a triangle. Then triangles TMP and PRQ are similar and give

$$\frac{t}{y} = \frac{e}{a},$$

i.e.,

$$t = y \cdot \frac{e}{a}.$$

So that

$$\mathrm{OT} = \mathrm{OM} - t = x - y \cdot \frac{e}{a}. \tag{6}$$

Since a/e is earlier determined, point T can now be located and joining T and P the tangent line at P can be obtained. For example, let us draw the tangent at (2, 1) to the ellipse

$$x^2 + 3y^2 = 7.$$

To find the value of a/e at (2, 1), we write

$$(2 + e)^2 + 3(1 + a)^2 = 7.$$

giving, on simplification and on neglecting e^2 and a^2,

$$4e + 6a = 0$$

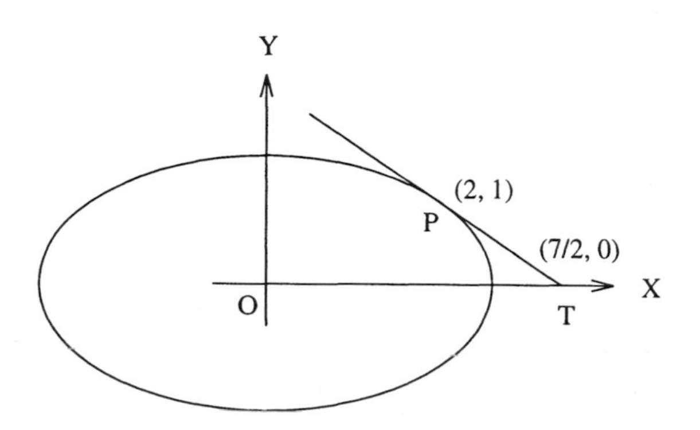

so that,

$$\frac{e}{a} = -\frac{3}{2}.$$

And as in (6),

$$\mathrm{OT} = \mathrm{OM} - t = 2 - 1 \cdot (-\frac{3}{2}).$$

So take T at point (7/2, 0) and join TP to get the tangent at P.

From Barrow's procedure to the present definition of a derivative was just a short distance. Let

$$y = f(x)$$

and let us apply Barrow's procedure to it. The first step would be

$$y + a = f(x + e),$$

i.e.,

$$f(x) + a = f(x + e)$$

so that

$$a = f(x + e) - f(x)$$

and

$$\frac{a}{e} = \frac{f(x+e) - f(x)}{e}.$$

When ultimately we set $e = 0$, we get the value of a/e. Apart from logical refinements following from the concept of limit, this is virtually the definition of a derivative of $f(x)$ at x.

5. Newton and Leibnitz.

It was at this stage that Leibnitz and Newton came on the scene: one a philosopher-mathematician and the other, a physicist-mathematician. It is not clear how much they got from the predecessors. Probably nothing: certainly not much. The problems which stirred their contemporaries like Fermat and Descartes and Barrow must have also been subjects of serious contemplation in the minds of Leibnitz and Newton. With better insights and better visions they must have been able to see farther than their other contemporaries. They must have vividly realized the possibilities the subject holds for mathematics as a whole and must have clearly foreseen the variety of applications the method may ultimately have in different areas – in problems of curves, of growths, of extreme values, and of physics in general.

Till Barrow's time, it was seen as a method which helps draw tangents. Even Barrow who showed an inclination to view it from a new aspect ultimately used the value of a/e to help draw tangents. They all looked at it from this limited point of view.

Even Leibnitz's first article published in 1684 was under the title

> "A new method for maxima and minima as well as tangents, which is impeded neither by fractional nor irrational quantities, and a remarkable type of calculus for them".

The title shows that Leibnitz prefers to indicate the roots of the calculus; but his last words, "and a *remarkable type of calculus* for this" are extremely refreshing. Leibnitz was a philosopher; he would not use terms loosely. These words are an index of his genuine assessment of the position of the subject then and the future development he envisaged for it. In this paper, Leibnitz gave a brief exposition of his *differential calculus* , the calculus of differentials. By establishing formulae such as

1. $\mathrm{d}(x^n) = nx^{n-1}\,\mathrm{d}x$, and
2. $\mathrm{d}(1/x^n) = -(n/x^{n+1})\,\mathrm{d}x$

and rules such as

1. $\mathrm{d}(uv) = u\,\mathrm{d}v + v\,\mathrm{d}u$, and
2. $\mathrm{d}(u/v) = (v\,\mathrm{d}u - u\,\mathrm{d}v)/v^2$,

he facilitated and made almost mechanical the work of calculating derivatives. The name Differential Calculus and the notation $\mathrm{d}y/\mathrm{d}x$, $\mathrm{d}^2y/\mathrm{d}x^2$, ...for derivatives is all a creation of Leibnitz.

Leibnitz concentrated his attention on one aspect of the subject, that of developing an ability to determine extreme values of all kinds of simple and non-simple functions and drawing tangents to all curves given by equations simple or otherwise. And since derivatives were needed for this purpose, he developed a calculus with formulae and rules with the help of which derivatives of more complicated functions could be calculated with speed and ease.

Newton looked at the subject more from the point of view of problems of physics. He sought and succeeded in obtaining a *meaning* for the derivative. His word for a variable was *fluent*, (that which flows, that which varies), and his word for a derivative was *fluxion*, meaning *rate of growth* of the fluent. His applications of fluxions to various problems of physics involving motion resulted from his rate-of-growth concept of a fluxion.

In his book, "Quadrature of curves", conceived much earlier but published in 1704, while determining the fluxion of x^3, Newton uses the following language:

> "In the same time that x, by growing, becomes $x + 0$, the power x^3 becomes $(x + 0)^3$, that is,
>
> $$x^3 + 3x^2 \cdot 0 + 3x \cdot 0^2 + 0^3$$
>
> and the *growths* in x and x^3, namely
>
> $$0 \quad \text{and} \quad 3x^2 \cdot 0 + 3x \cdot 0^2 + 0^3$$

are to each other as

$$1 \text{ is to } 3x^2 + 3x \cdot 0 + 0^2.''$$

Letting increment 0 vanish, he concludes that

"The *rate of change* of x^3 with respect to x is $3x^2$."

Newton uses 0 with the same meaning as the e which was used above in explaining the work of Fermat and Barrow. But the change of notation is not the point. The change of *outlook* is what mattered. It gave the result a *meaning* and a *general setting.* What was important for mathematics was this *new outlook* – that a *fluxion* (or *derivative*, by whatever name you call it) is a *rate of change.* It is this new outlook which set the wheels moving, in so far as Newton's ideas were concerned. It made the subject grow into a *new calculus*, so to speak, of *rates of growth.*

Newton, in his work, considered two types of problems. In one, he obtained relations between fluxions from relations between fluents – the problem of differentiation as we now call it. And, in the second, he gave methods to obtain relations between fluents if relations between fluxions were given – the problem of solving *differential equations.* Newton's contribution on all counts, appears to be broader, with applications emphasized more than devices, rules or symbolism to make calculations of fluxions simpler. In a way, the work of Leibnitz and that of Newton, from this point of view, were mutually complementary.

And thus, what started as a method, ultimately grew into a *calculus*, a calculus which, in a way, changed the whole face and form of Mathematics. It marked the end of one era and beginning of another.

Exercises

1. Use Apollonius' method and obtain values of sub-normal and sub-tangent at (1, 2) on the ellipse $3x^2 + 2y^2 = 11$.

2. Use Apollonius' method and obtain the value of the sub-normal at point (3, 5) on the hyperbola $4x^2 - y^2 = 11$.

3. Use the method of Roberval and locate the tangent at a point on the ellipse $2x^2 + y^2 = 8$.

4. Use Descartes' method and locate the normal to the ellipse $x^2 + 2y^2 = 9$ at the point (1, 2) on it.

5. Find the sub-tangent at the point (1, 4) on the parabola $y^2 = 16x$, using Fermat's method.

6. Use the method of Fermat and find the maximum/ minimum of the function $3x^4 - 16x^3 + 6x^2 - 48x$.

7. Find the value of (a/e) in Barrow's notation at the point (-3, 2) on the ellipse $x^2 + 3y^2 = 21$.

8. Find the derivative of $x^3 + 2x^2 - 7$ by Newton's method.

9. Using Newton's method, find the derivative of $(1/x)$.

10. Find derivative of $x^3 - 5x^2 + 7x$ by Newton's method and show that this derivative is zero at $x = 1$, and also at $x = (7/3)$.

Further Reading

1. J. F. Scott, *A History of Mathematics*, Taylor and Francis Ltd., London, 1969.

2. Carl B. Boyer, *The History of the Calculus and its Conceptual Development*, Dover, New-York, 1959.

CHAPTER SIX

THE FIRST NON-COMMUTATIVE ALGEBRA

1. The first great crisis in mathematics.

The appearance of numbers – rational, irrational, complex - is so common in mathematics to-day that one finds it difficult to believe that each one of the new kinds of numbers, on their first entry had created a crisis and a challenge in the development of mathematics. Negative numbers were not easily accepted; nor were irrational and complex numbers. Each type produced a crisis: but every time it happened, mathematicians showed enough ingenuity to face the challenges and resolve the crises. And every time the crisis was resolved, mathematics did not falter but advanced further.

The first such crisis was created by irrational numbers such as $\sqrt{2}$ and $\sqrt{5}$. It was the fond conviction of the Pythagorean school that all numbers can be ultimately expressed in terms of natural numbers.

This belief of the Pythagoreans that all numbers ought to be expressible as rational numbers meant, in geometry, that every two lines must be commensurable in length. And when it was clear that a side of a square and its diagonal are not commensurable, they no longer could justifiably assume that every pair of given lines was commensurable. For the Pythagoreans, this was extremely scandalous. They had already given proofs of geometrical theorems assuming that commensurability unfailingly existed. This was a terrible shake-up of their prestige. The crisis was too scandalous for the Pythogoneans to withstand and could be one of the psychological reasons for the ultimate break-up of the school.

This crisis in geometry was resolved later by the Greek geometer Eudoxus with his novel way of handling the concept of proportion. In fact, one wonders if this theory of proportion would ever have been advanced at so early a stage in mathematics had not this crisis of incommensurability not then propped up.

The arithmetical crisis involved in respect of irrational numbers such as $\sqrt{2}$, was however, resolved very much later, after about two thousand years when Weierstrass, Dedekind and others made a concerted effort for the Arithmetization of Analysis. And it was remarkable that as a result of this movement of theirs, and the results that consequently came up, Pythogoneans were completely vindicated and irrational numbers were fully constructed through the use of natural numbers only.

2. The second great crisis.

The second great crisis arose almost two thousand years later; it was in respect of *complex numbers.* It was in the work of the Italian mathematician Cardano that complex numbers first appeared. Cardano himself was not happy with them. Even when he felt sure that they must be discarded, he was caught in a kind of trap and he found himself helpless. He was not ready to accept them as numbers: but when he worked with them and applied to them the usual ordinary rules of real numbers, he found, to his utter surprise that they led to correct results.

The unexpected result came to his notice when he was in the process of discovering his method of *solving a cubic equation.* He first considered the equation

$$x^3 + 6x = 20$$

To solve this, he set $x = u - v$ and got the original equation transformed to

$$(u - v)^3 + 6(u - v) = 20$$

That is,

$$u^3 - v^3 - 3uv(u - v) + 6(u - v) = 20 \tag{1}$$

In this he put

$$3uv = 6 \tag{2}$$

$$\text{so that} \quad u^3 - v^3 = 20 \tag{3}$$

and argued that for values of u and v satisfying these two latter equations, $x = u - v$ satisfies equation (1) and consequently, the original equation $x^3 + 6x = 20$. Calculating u^3v^3 from (2) and using this value of u^3v^3 along with the equation (3), he obtained the value of $u^3 + v^3$, and got

$$u^3 = \sqrt{108} + 10 \quad \text{and} \quad v^3 = \sqrt{108} - 10$$

so that, the roots of $x^3 + 6x = 20$ turned out to be

$$x = \sqrt[3]{\sqrt{108} + 10} - \sqrt[3]{\sqrt{108} - 10}$$

Algebraists at Cardano's time were not unfamiliar with surds. Cardano worked out the two cube-roots concerned as $\sqrt{3} + 1$ and $\sqrt{3} - 1$; so that one root of the cubic equation $x^3 + 6x = 20$ was determined to be

$$x = (\sqrt{3} + 1) - (\sqrt{3} - 1) = 2.$$

The final result surprised him; but also pleased him. He realized that formal work, whether justifiable or otherwise, could lead to a correct conclusion. This realization – though he never put it in so many words – gave him confidence in dealing with the more mystifying situation which he came across when he next tried to find a solution of

$$x^3 = 15x + 4$$

to solve which he put $x = u + v$ and followed precisely the same procedure as above to finally get

$$x = \sqrt[3]{2 + \sqrt{-121}} + \sqrt[3]{2 - \sqrt{-121}}$$

Undaunted by the negative number -121, appearing under the square-root sign, he applied to the expression, the ordinary rules of algebra and obtained the two cube-roots as

$$2 + \sqrt{-1} \quad \text{and} \quad 2 - \sqrt{-1}$$

and adding them got

$$x = 4$$

as the value of x. On verification, he found that $x = 4$ was, indeed, a root of $x^3 = 15x + 4$.

The final gain of all this non-challent work of Cardano was:

Complex Numbers Came To Stay for Good.

But not immediately. Cardano's work was of the year 1545. Since, solution of a cubic equation was of interest to many, this work of Cardano came to the notice of almost every one who was then working in Algebra. But other algebraists of the period did not take kindly to these imaginary numbers. They could not make themselves bold enough for the purpose: they would neither accept them nor reject them. They simply kept on dodging the situations in which they were likely to confront them; until after almost hundred years, in 1637, Descartes introduced them in his *La Geometry.* After Descartes, other mathematicians took to their use more freely. Descartes had used for them the name *Imaginary Numbers.* It was much later, in the year 1829 that Gauss gave them the name *Complex Numbers.*

In the intervening years, a geometrical representation for complex numbers as points in the coordinate plane was suggested by Wassel in 1797 and Argand in 1806. Euler used them extremely freely and secured a central place for them in the body of mathematics by establishing the fundamental intra-functional relation

$$e^{ix} = \cos x + i \sin x$$

and using such relations freely for deriving a number of other more advanced results of the same nature.

In spite of all this free use and in spite of the kind of reality which the geometrical representation brought to them, people could not easily digest the *queer hybrid* look of the number $a + bi$, a part of which is a *real* number and another, which even after all was said and done, continued to be designated *imaginary.* It was William Rowan Hamilton, who first removed this veil of mysteriousness which surrounded the complex numbers $a + bi$ by giving to the set of complex numbers the following straight forward definition as

> A set of *ordered number pairs* (a, b) of real numbers in which
>
> 1. $(a, b) = (c, d)$ if and only if $a = c$ and $b = d$
>
> 2. $(a, b) + (c, d) = (a + c, b + d)$ and
>
> 3. $(a, b)(c, d) = (ac - bd, ad + bc)$.

This was the year 1833. Not only did this dispel all mysteriousness from complex numbers : it also proved providential in the later efforts of Weierstrass and others to build up, stage by stage, all numbers from natural numbers and thus help realize the old cherished dream of the Pythagoreans.

3. The number-pair episode revisited.

In 1833 Hamilton published a paper entitled "The Algebra of Couples". In this paper he starts with a set consisting of ordered number-pairs of real numbers and in this set defined equality, and two operations which he calls *Addition* and *Multiplication* modeled on the lines of those in Complex Numbers. Thus, since, with complex numbers

$$(a + bi)(c + di) = (ac - bd) + (ad + bc)i$$

he defined the *product* for two couples (a, b) and (c, d) as the couple

$$(ac - bd, ad + bc)$$

He has, at no place in the paper concerned, even hinted that his couple is a representation of a complex number. But his contemporaries and successors, already in search for a clear, straight non-mysterious representation of complex numbers, found Hamilton's couple just the thing they were in search of.

It is true that Hamilton provided them this number-pair representation of complex numbers. But it was incidental. That was not

Hamilton's interest. The fact that he followed his paper on the algebra of couples by an attempted algebra of triplets - unsuccessful though the attempt was - and still later by the successful attempt at building up an algebra of quadruples, fairly clearly indicates the direction and emphasis of Hamilton's research in this area.

All the same, the credit of giving the number-pair form to a complex number, must surely go to Hamilton.

4. Hamilton's discovery of the Quaternions.

Hamilton must have been pleased with his algebra of couples. And it must have struck him that if a useful algebra could be so developed out of number-pairs (a, b), may it not become possible to develop an algebra of number-triplets (a, b, c), and may it not prove to be a useful algebra playing a role in three dimensions parallel to the role complex numbers were playing in two dimensions?

For such an algebra of triplets to be possible, one of the first requirements would be an additional *basis element.* This was simple enough. The new basis element introduced by Hamilton was named j with $j^2 = -1$. Thus the three basis elements for the triplet algebra were visualized to be

$$1, \ i, \ j$$

so that a triplet (a, b, c) could now be written

$$a + bi + cj$$

Equality and addition were defined in a natural way as

$$(a, b, c) = (d, e, f) \text{ if and only if } a = d, b = e, c = f$$

and

$$(a, b, c) + (d, e, f) = (a + d, b + e, c + f).$$

For an algebra to be possible in any set, the operations have to be so defined that the set remains *closed* under those operations. The operation of addition, as defined above, satisfies the requirement in the sense that a sum of two triplets is again a triplet.

This, however, did not happen when Hamilton multiplied triplet $(a+bi+cj)$ by triplet $(d+ei+fj)$. The product turned out to be

$$(ad-be-cf)+(ae+bd)i+(af+cd)j+(bf+ce)ij$$

a triplet with an additional term involving ij. This product could become a triplet if ij were zero. But to take $ij=0$ when $i^2=-1$ and $j^2=-1$ appeared odd and uncomfortable to Hamilton. As a next best step he decided to try the rather unorthodox hypothesis that

$$ji=-ij$$

and put each equal to k where k^2 also would now be -1. Under this hypothesis, he wrote the product again as

$$(ad-be-cf)+(ae+bd)i+(af+cd)j+(bf-ce)k \qquad (4)$$

This gave rise to another problem.

Hamilton had observed that in the algebra of couples, where

$$(a+bi)(c+di)=(ac-bd)+(ad+bc)i$$

the square

$$(ac-bd)^2+(ad+bc)^2$$

of the modulus of the product, is equal to the product

$$(a^2+b^2)(c^2+d^2)$$

of the squares of the moduli of the factors.

He had called it the *law of the moduli* and he expected the algebra of triplets to satisfy this law. When he inspected the product (4) from this point of view, he did find that the square

$$(ad-be-cf)^2+(ae+bd)^2+(af+cd)^2+(bf-ce)^2$$

of the modulus of the product came out to be equal to the product

$$(a^2+b^2+c^2)(d^2+e^2+f^2)$$

of the squares of the moduli of the factors.

The enigma was now deeper still. If k is set equal to 0, the triplets are closed under multiplication, but the law of the moduli is not satisfied; and if k is not set equal to 0, the law of the moduli is satisfied but triplets do not remain a closed set under multiplication.

The difficulty was a major one : almost insurmountable. Admitting k would mean admitting a fourth dimension for multiplication of triplets. And a fourth dimension could not even be imagined. As with complex numbers, Hamilton was looking forward to applications of triplets to rotation problems in three dimensions. And space has no fourth dimension. And again if the product is four-dimensional the factors also would have had to be four-dimensional. So the product would go beyond the algebra of triplets. It would be an algebra of quadruples.

Should he then give up the attempt to build up an algebra of triplets and attempt one of four dimensions?

Is there no way out?

And Hamilton in a flash, took a revolutionary leap forward and the entire problem got a new look and a new direction. In a letter to John Graves, a mathematician friend of his, sent on October 17, the day after his final decision and discovery of quaternions, Hamilton writes:

> And here, there dawned on me the notion that we must admit, in some sense, a *fourth dimension* of space for the purpose of calculating with triplets ... or transferring the paradox to algebra, we must admit a third distinct imaginary symbol k, not to be confused with either i or j, but equal to the product of the first as multiplier, the second as multiplicand, and therefore I was led to introduce *quaternions* such as $a + bi + cj + dk$ or (a, b, c, d).

The entry which he made in his *note-book* on the very night of the discovery is more clear:

> I believe I now remember the order of my thought. The equation $ij = 0$ was recommended by the circumstance

that

$$(ax - y^2 - z^2)^2 + (a + x)^2(y^2 + z^2) =$$

$$(a^2 + y^2 + z^2)(x^2 + y^2 + z^2)$$

I therefore tried whether it might not be true that

$$(a^2 + b^2 + c^2)(x^2 + y^2 + z^2) =$$

$$(ax - by - cz)^2 + (ay + bx)^2 + (az + cx)^2$$

but found that the equation required, in order to make it true, the addition of $(bz - cy)^2$ to the second member. This *forced* on me the non-neglect of ij and *suggested* that it might be equal to k, a new imaginary.

In a letter to his son Archibald written twenty-two years later, in 1865, he recounts very vividly the personal happenings of that eventful evening of October 16, 1843:

Every morning in the early part of the above-cited month, on my coming down to breakfast, your brother William Edward and yourself used to ask me: "Well, Papa, can you multiply triplets?" where to I was always obliged to reply, with a sad shake of my head, "No, I can only add and subtract them . . . ."
But on the 16th day of the month, which happened to be a Monday and a Council day of the Royal Irish Academy, I was walking in to attend and preside, and your mother was walking with me, along the Royal Canal . . . ; and although she talked with me, now and then, yet an undercurrent of thought was going on in my mind, which gave at last a result, whereof . . . I felt at once the importance. An electric circuit seemed to close, and a spark flashed forth, the herald (as I foresaw immediately) of many long years to come of definitely directed thought and work.

> I pulled out on the spot a pocket-book, which still exists, and made an entry there and then. Nor could I resist the impulse – unphilosophical as it may have been – to cut with a knife on a stone of Broughham Bridge, as we passed it, the fundamental formula with the symbols $i,\ j,\ k$:
>
> $$i^2 = j^2 = k^2 = ijk = -1$$
>
> which contains the solution of the problem; but of course, as an inscription has long since moldered away.

The cutting on the stone had indeed gone. Even the stone which carried it had been lost. But the Irish people had not so easily forgotten their hero nor the melodramatic way in which one of their brilliant sons had made the first discovery of a *non-commutative algebra.* The stone is gone. Today, a cement tablet embedded in a stone of the bridge proudly tells the story.

Here as he walked by on the 16th of October 1843 **Sir William Rowen Hamilton** in a flash of genius discovered the fundamental formula for quaternion multiplication $i^2 = j^2 = k^2 = ijk = -1$ and cut it in a stone of the bridge.

5. The Quaternion Algebra.

A quaternion is an ordered quadruple (a, b, c, d) of real numbers. To build up an algebra with quaternions as elements, a relation of equality and operations of addition and multiplication are defined as under.

1. $(a, b, c, d) = (e, f, g, h)$ if and only if $a = e$, $b = f$, $c = g$ and $d = h$.

2. $(a, b, c, d) + (e, f, g, h) = (a + e, b + f, c + g, d + h)$

3. $(a, b, c, d)(e, f, g, h) = (ae - bf - cg - dh, af + be + ch - dg, ag + ce + df - bh, ah + de + bg - cf)$

It can be seen from these definitions, that quaternions are *closed* under addition and multiplication. It also can be verified directly that the product

$$(a^2 + b^2 + c^2 + d^2)(e^2 + f^2 + g^2 + h^2)$$

of the squares of the moduli of the factors, is equal to

$$(ae - bf - cg - dh)^2 + (af + be + ch - dg)^2 +$$

$$(ag + ce + df - bh)^2 + (ah + de + bg - cf)^2$$

the square of the modulus of the product; which means that the Law of the Moduli is true for quaternions as it was for couples.

The Quaternion Algebra is, indeed, an extension of Real and Complex algebras. To see this consider the special quaternions of the form $(a, 0, 0, 0)$ and observe that

$$(a, 0, 0, 0) + (e, 0, 0, 0) = (a + e, 0, 0, 0)$$

and

$$(a, 0, 0, 0)(e, 0, 0, 0) = (ae, 0, 0, 0)$$

so that if a quaternion of the form $(a, 0, 0, 0)$ is identified with the real numbers a, we see that the Real Algebra gets embedded in the Quaternion Algebra.

Similarly, if we restrict our attention to quaternions of the form $(a, b, 0, 0)$ we find that

$$(a, b, 0, 0) + (e, f, 0, 0) = (a + e, b + f, 0, 0)$$

and

$$(a, b, 0, 0)(e, f, 0, 0) = (ae - bf, af + be, 0, 0)$$

and if quaternions of the form $(a, b, 0, 0)$ are identified with complex numbers of form (a, b), we conclude that the Complex Algebra is embedded in Quaternion Algebra.

It is true in complex numbers, that, in form, (a, b) is more elegant than $a + bi$. Yet on the operative side, it is easier to multiply by $a + bi$ as if it is an ordinary algebraic multiplication and replace $i^2 = -1$ wherever it occurs rather than remember the rule $(a, b)(c, d) = (ac - bd, ad + bc)$. With quaternions, the rule for multiplication is still more cumbersome. The way out is found by defining four special quaternions, 1, i, j, k as under:

$$\begin{aligned} 1 &= (1, 0, 0, 0) \\ i &= (0, 1, 0, 0) \\ j &= (0, 0, 1, 0) \\ \text{and} \quad k &= (0, 0, 0, 1) \end{aligned}$$

This firstly makes it possible to write a quaternion (a, b, c, d) as

$$a + bi + cj + dk.$$

Next, by using definitions of addition and multiplication, one can establish the results

(a) $i^2 = j^2 = k^2 = ijk = -1$

the fundamental formula of quaternion multiplication

(b) $(ij)k = i(jk)$

showing that the quaternion algebra is *associative*

and (c) $ij = -ji$

showing that the quaternion algebra is *non-commutative.*

Hamilton had discovered the first non-commutative algebra. This was in 1843. In 1857 came the second: namely, Cayley's *matrix algebra.*

It is true that Hamilton created the first non-commutative algebra. But that was not the object of his search. In 1833, he had created an algebra of couples taking cue from the algebra of complex numbers and in the process had come across a few interesting results, one of which was the result

$$(a^2 + b^2)(c^2 + d^2) = (ac - bd)^2 + (ad + bc)^2$$

from the number-theoretic area. He naturally hoped to discover more of similar results and with this desire attempted the construction of an algebra of triplets. He failed in this latter attempt and, fortunately for mathematics, did not give up the effort altogether but jumped on to building an algebra of ordered quadruples and *succeeded* in the same. The algebra of quaternions was created. And this algebra happened to be non-commutative and also proved to be the *first* of its kind.

The *first non-commutative algebra* was discovered: It was discovered on Monday, October 16, 1843. It was created by

WILLIAM ROWAN HAMILTON.

Exercises

1. Use Cardano's method and solve the equation

$$x^3 = 30x + 36$$

2. Use the result $(a^2 + b^2)(c^2 + d^2) = (ac - bd)^2 + (ad + bc)^2$ and express 17×65 as a sum of two squares in as many ways as can be done.

3. In 1843 December, Mr. J. T. Graves, a friend of Hamilton created an *Octonian Algebra* of 8 basis elements

$$1, \ i, \ j, \ k, \ l, \ m, \ n, \ o$$

for which he gave the multiplication table

   (a) Use this table and find the product of the octonions

   $$(2 + i + 3l - m + 2o) \text{ and } (3 + 4j - l + 2m + n - 4o)$$

   and express the product 19×47 as a sum of 8 squares thus verifying the Law of Moduli for octonions.

   (b) Show that the Octonion Algebra is non-commutative and that it is also not associative.

	i	j	k	l	m	n	o
i	-1	k	$-j$	m	$-l$	$-o$	n
j	$-k$	-1	i	n	o	$-l$	$-m$
k	j	$-i$	-1	o	$-n$	m	$-l$
l	$-m$	$-n$	$-o$	-1	i	j	k
m	l	$-o$	n	$-i$	-1	$-k$	j
n	o	l	$-m$	$-j$	k	-1	$-i$
o	$-n$	m	l	$-k$	$-j$	i	-1

4. Elements a, b, c, d and s, t, u, v are real numbers. A and S are matrices

$$\begin{pmatrix} a & b \\ c & d \end{pmatrix} \text{ and } \begin{pmatrix} s & t \\ u & v \end{pmatrix}$$

respectively, of order 2 each. The sum of all such matrices, denoted by +, is defined by

$$\mathrm{A} + \mathrm{S} = \begin{pmatrix} a+s & b+t \\ c+u & d+v \end{pmatrix}.$$

Their Cayley product denoted by AS is the matrix

$$\begin{pmatrix} as+bu & at+bv \\ cs+du & ct+dv \end{pmatrix}.$$

Show that

(a) Cayley multiplication is not commutative

(b) but that it is associative.

5. Let A and S be matrices as above, their sum defined the same way, and let AS and SA be their Cayley products in the two orders. Let a new multiplication, denoted by $\times$, be defined by

$$\mathrm{A} \times \mathrm{S} = \frac{\mathrm{AS} + \mathrm{SA}}{2}$$

Show that

(a) operation $\times$ is commutative

(b) but that it is not associative.

6. Let A and S be matrices as above, their sum defined the same way, and let AS and SA their Cayley products. Let a new operation, denoted by *, be defined by

$$A * S = AS - SA$$

Show that the operation is neither commutative nor associative.

7. Use Quaternion Law of Moduli and express 39×91 as a sum of four squares in as many ways as can be done.

Further Reading

1. Howard Eves, *Great Moments in Mathematics After 1650*, Mathematical Association of America, 1978.

2. B. L. van der Waerden, *A History of Algebra From Al-Khwarizm to Emmy Noether*, Springer-Verlag, 1985.

3. *The Mathematical Papers of Sir William Rowan Hamilton*, Vol. III, Cambrige University Press, 1967.

CHAPTER SEVEN

ARITHMETIZATION OF ANALYSIS

1. Imperatives for the growth of Mathematics.

Man is, by nature, inquisitive, curious to know the unknown, to solve problems that come in his way, to face and overcome the challenges that face him. This is how Science grows.

There are also, by and large, the imperatives that guide the growth of Mathematics. Challenging problems and well-supported feasible looking conjectures have always provided an added impulse for mathematicians to add to their discipline, sometimes substantially.

We have seen one such instance; the problem of Pappus. A problem of the Greek era, which had defeated all efforts of the able geometers of the period like Euclid and Apollonius, was picked up as a *challenge* by a later seventeenth century French Philosopher-mathematician Rene Descartes. Endowed with an exceptional insight and imagination, Descartes could see that the great of the Greek time failed because their method was weak; and that the solution could be found, if found at all, only if the problem could be attacked by some *new method.* Descartes discovered one such method, namely the method of Analytic Geometry and thereby created a *new branch of mathematics.*

Like problems, *conjectures* have also proved to be powerful driving forces for the creation of Mathematics. *Fermat's last theorem* was one such; probably, the most fruitful in respect of the amount of mathematics it gave rise to.

The story of this conjecture is quite interesting. In 1621, one *Bachet De Meziniac* published a French translation of the book *Arithmetica* of Diophantus, a Greek number-theorist. This book poses a number of problems about numbers, mostly about natural numbers. Problem 8 in the book reads:

> "To divide a square-number as a sum of two square-numbers."

To do this is easy enough. What is important in this regard, is a marginal note which Fermat made in his personal copy of Bachet's book which reads:

> "To divide a cube into two cubes, a fourth or in general any power whatever, into two powers of the same denomination above the second is impossible, and *I have assuredly found an admirable proof of this*, but the margin is too narrow to contain it."

In other words, he asserted two things:

> If x, y, z be two non-zero integers, and $n > 2$, then z^n cannot be expressed as $x^n + y^n$

and that

> he had discovered an admirable proof of it.

Normally, marginal remarks are taken as casual notes. But they could not be so construed here. Because, Fermat was no ordinary mortal. A versatile mathematician as he was, he had given ample evidence of his originality and exceptional capacity to discover new results and create new methods.

But Fermat, even later, gave no idea of what this admirable proof was which, he professed, he had possessed. The said marginal remarks came to the notice of the others only after Fermat's death. A thorough search for the said proof was undertaken amongst whatever papers, notes and letters he had left behind. But to no avail.

The said original admirable proof was not to be traced; but Fermat's remarks about his discovery of the same were quite assertive; not to be so easily ignored. Contemporary and succeeding mathematicians had no alternative but to conclude that the statement about $x^n + y^n = z^n, n > 2$ was true and further to believe that a proof of it existed and that Fermat had found it, but that now it was missing. They therefore gave up the search for it in Fermat's papers and took a decision to discover it themselves, since such a proof had existed and Fermat had had it. First rate number-theorists such as Euler, Kummer and others spent a lot of their time, energy and ingenuity in this effort. Their efforts failed. But, in a way, not completely. It is true that they could not get a proof of what since is called Fermat's Last Theorem. But in the process, they created a lot of important Number-Theory, showing how a good conjecture made by an able knowledgeable mathematician becomes the occasion for a valuable advancement of a theory of mathematics. Fermat's Last Theorem is not the only example of this. There are other instances also. But this one, even singularly, is surely enough to show how good conjectures like good problems promote good mathematics.

Just as problems and conjectures appear in mathematics periodically, so also do crises : logical and fundamental. They arise now and again. They usually arise because when a new fruitful method to derive new results is discovered, not only the discoverer but even others are so enthusiastic about it and the possibility of discovering new results so tempting that nobody has either time or patience to care for the various pros and cons of the method or for the conditions and restrictions under which the new derivations are or are not valid. Generally, even the discoverer of the new method is not aware of these conditions or has no patience even to think a second time about the validity or otherwise of his presumptions.

This was, for example, what had happened in the past with the Pythagoreans who believed that every two line-segments were mutually commensurable and on this belief, gave proofs of a number of theorems of geometry. And when they found that their own work led them to the fact that a side of a square is not commensurable to its diagonal,all their proofs collapsed and fell apart. Fortunately, Eudoxus came forward with a new ingenious theory of proportions and saved the theorems.

But there was another intriguing and embarrassing point involved in the discovery: it was the appearance of an *irrational* number $\sqrt{2}$. Pythagoreans had a conviction that all numbers could be constructed out of natural numbers. But they found that $\sqrt{2}$ could not be so constructed. Is $\sqrt{2}$ then a number? But a line of that length could be constructed. So it is a number. The crisis was not as simple as it appeared in the beginning. Neither the Pythagoreans could resolve it nor any other Greeks. This crisis could be resolved only after about two thousand years, when mathematicians of the time of Dedekind and Cantor could give a clear unimpeachable definition of real numbers based on rational numbers. Till then, the fate of numbers like $\sqrt{2}$ and π and e had to be kept in the hanging.

A number of such crises have arisen in Mathematics. Each time they arose, mathematicians and logicians had to think hard to resolve them and cleanse mathematics of the possible blemishes and help it rise and regain for it the supreme position, it now has, of a logically precise discipline.

2. The crisis in Calculus.

Such a crisis arose after the birth of Calculus. The story of how mathematicians dealt with it and ultimately resolved it, is indeed, a highly illuminating story and deserves to be told and continuously re-told; and read and continuously re-read.

The crisis concerned arose almost immediately after Calculus was born. Newton published his *Quadrature of Curves* in 1704. In this book Newton argues as below in showing that the derivative of x^3 is $3x^2$. We have given this argument verbatim in the chapter on Differential Calculus. We repeat it here for ready reference.

Without clearly mentioning what his symbol *o* stands for, he argues:

> In the same time that x, by growing, becomes $x + \circ$, the power x^3 becomes $(x + \circ)^3$, that is,
>
> $$x^3 + 3x^2 \cdot \circ + 3x \cdot \circ^2 + \circ^3$$
>
> and the *growths* in x and x^3, namely
>
> $$\circ \quad \text{and} \quad 3x^2 \cdot \circ + 3x \cdot \circ^2 + \circ^3$$
>
> are to each other as
>
> $$1 \quad \text{is to} \quad 3x^2 + 3x \cdot \circ + \circ^2.$$
>
> Now, let the basic increment $\circ$ vanish, so that the previous proportion of their increments, becomes as

$$1 \text{ is to } 3x^2$$

wherence, the rate of change of x^3 with respect to x is

$$3x^2.$$

The fault in the argument is more than obvious. Firstly, he does not specify what o stands for. And then at one stage in the argument he insists that o is not zero and therefore one may validly divide by it; and at a later stage he considers o to be zero and ignores it from further argument. This is plainly a *change of hypothesis* in the middle of an argument. When it is so plain even to a casual reader of the argument, it is no wonder that a contemporary English meta-physicist, Bishop Berkley, publicly made such a charge and objected to it. Now Berkley was a meta-physicist of a good standing and reputation : and the charge he made was serious. It shook the very foundation on which Calculus stood.

On the one hand, the charge was serious and mathematicians could not give any satisfactory reply to it. They either evaded it or gave replies which were just ridiculous. The reply of Johanne Bernoulli that

> a quantity which is increased or decreased by an infinitely small quantity is neither increased nor decreased

is typical of the replies mathematicians could venture to advance to meet the charge of Bishop Berkley. Even Bernoulli in his quieter moments later, must have realized how flippant this reply of his was. But he had no time to wait and take a more balanced position in the matter and give a suitable reply. Nor, perhaps, he had even the ability to do it. The case with other practicing mathematicians of that time was not any different.

The crisis was not just this. Even granting that the argument which lay at the foundation of Calculus was faulty and brittle, the surprising picture, on the other side, was the fact that results based on it were remarkably sound. A number of good new results were daily being derived : a number of problems were getting correctly solved : a number of striking results about the usual functions were coming to the fore, unifying functions of different origins, different definitions and different looks such as $\sin x$, $\arctan x$, e^x, $\log(1+x)$, or $(1+x)^{p/q}$ — unifying all these in one common form, namely, that of Power Series, thus bringing to light a fresh and pleasant general relationship least suspected till then.

This was really remarkable. Because, most of the above results were derived just *formally* with no particular attention to conditions of validity : almost non-challently done. And yet, so many of the results were good : looked perfect : proved useful. Perhaps they were arguing to themselves that the conditions put on the theorems, even when they were explicitly stated, had been far too restrictive : were yet to be refined and that some one who will come later would refine them and indirectly vindicate the position they had been taking. Such things, they said to themselves, had happened in the past. Had not a similar situation once disturbed Cardano when he came across the form

$$\sqrt[3]{2+\sqrt{-121}}+\sqrt[3]{2-\sqrt{-121}}$$

while solving the cubic equation $x^3 = 15x + 4$? Cardano was really not at all happy with the expression $\sqrt{-121}$, the square-root of a negative number which, by no stretch of imagination could be considered a valid mathematical entity. But gambler by temperament as well as by profession, that he was, Cardano simply gambled with the expression $\sqrt{-121}$, applied the rules of real algebra to the expression which he had got for a root of $x^3 = 15x + 4$, found the cube-roots concerned, namely $2+\sqrt{-1}$ and $2-\sqrt{-1}$, and discovered to his utter surprise that their sum 4 was really a *real* root of $x^3 = 15x + 4$.

The gamble had worked. And the unexpected success gave him and some of his contemporaries a kind of confidence to work with entities like $\sqrt{-121}$. They continued to work with them, with increasing acceptance of the same by the rest. The precise logical setting for it came much later, after almost three hundred years, at the hands of Hamilton.

This seems to be the psychology of initial workers in a new area. They take chances; some chances work, some do not. This is also what happened in respect of Calculus. A number of results were derived; some good and some not so good. Those which were good were remarkably good; those which were not so good were almost absurd, unbelievably absurd. Even workers of the caliber of the great Euler did not deter from accepting some obviously absurd looking results and granting some sort of respectability to them. One such result which Euler considered acceptable, was

$$-1 = 1 + 2 + 4 + 8 + \cdots$$

obtained by him from the Binomial expansion

$$(1 - x)^{-1} = 1 + x + x^2 + x^3 + \cdots$$

by putting $x = 2$ in it, a value of x for which the expansion is not valid. The infinite series for $(1 - x)^{-1}$, in fact *does not converge* for x if $|x| \geq 1$; but at Euler's time, though the form of expansion was quite well known, the importance of the *concept of convergence* and the *test thereof* were not as firmly established and respected as they deserved. Look at another of his result also flowing from the same lack of insistence on conditions under which these new results hold and blindly doing the formal work. Using the same Binomial theorem, Euler got the two results

$$x + x^2 + x^3 + \cdots = \frac{x}{1 - x}$$

and

$$1 + \frac{1}{x} + \frac{1}{x^2} + \frac{1}{x^3} + \cdots = \frac{x}{x - 1};$$

the first is correct only if $|x| < 1$ and the second only if $|x| > 1$. Disregarding these conditions which are necessary for the validity of the two results, and taking it that both are correct for the same x, Euler added the two series and obtained the result

$$1 + \frac{1}{x} + \frac{1}{x^2} + \frac{1}{x^3} + \cdots + x + x^2 + x^3 + \cdots = 0.$$

It is surprising that such results were permitted to appear in published literature. Probably, it was considered acceptable under the compelling pressure of the *large number of good results* that mathematicians of the time had obtained by the use of the same methods. This fact had doubled and redoubled their confidence in the validity of their practices.

In a way, it is surprising that mathematicians of the seventeenth and the eighteenth centuries should have shown so much ignorance and so little respect for conditions under which given theorems are valid, when one remembers that they were all thoroughly drilled in the strict logical discipline of Euclid's geometry where every theorem is a *conditional truth.* When we notice this callousness in respect of the usual accompaniments of theorems, how much thought could we expect them to have paid to the deeper questions in regard to infinite processes such as, for example, whether the operation of addition which is commutative and associative for *finite* sums, continues to be so for infinite sums. Had they shown any awareness in suspecting the associativity of addition in an infinite series, they may not have thought that the sum

$$S_1 = 1 - 1 + 1 - 1 + 1 - 1 + - \cdots$$

is the same as the sum

$$S_2 = 1 - (1 - 1) - (1 - 1) - - \cdots$$

obtained from the first by associating the terms of the sum pairwise in one way, and the sum

$$S_3 = (1 - 1) + (1 - 1) + (1 - 1) + + \cdots$$

obtained by associating the terms pairwise in another way. They even argued that since $S_2 = 1$ and $S_3 = 0$, it would be quite proper to take the sum of S_1 to be $1/2$, the average of S_2 and S_3.

3. The awareness.

With time and increasing accumulation of such results, more and more mathematicians started realizing that justification for such *formal* work needs to be urgently and carefully looked into. They were not immediately worried about logical errors such as those pointed out by the meta-physicist, Bishop Berkley. They could afford to ignore it for some time and wait for other logicians to suggest a correction thereto or for a mathematician to give a *correct definition* of a *limit.* But the glaring absurd results that followed in its wake left the later mathematicians extremely uneasy.

The first mathematician to put the case squarely before the whole body of mathematicians was D'Alembert. D'Alembert was for a long time *secretair perpetuel* of the French Academy and, with Dennis Diderot, the moving spirit of the famous *Encyclopedie ou Dictionaire* of sciences (28 volumes, Paris, 1751-1772). D'Alembert wrote a number of articles in this *encyclopedie* and in the article entitled *differentel*, (vol. 4, 1754), he came to the expression for the derivative as the limit of a quotient of increments, that is, to what we now write as

$$\frac{dy}{dx} = \lim \frac{\delta y}{\delta x} \quad \text{as} \quad \delta x \to 0.$$

This leading idea, however, was not followed up immediately, either by D'Alembert himself or by others – probably because being an article in an encyclopedia, it might not have reached many : and to those whom it reached, the words *tends to zero* in it were suggestive of motion and reminded them of Zeno's paradox about the impossibility of motion.

A year later, in 1755, Euler published his *Institutiones Calculi Differentials.* He wrote

> To those who ask what the infinitely small quantity in mathematics is, we answer that it is actually 0. Hence there are not so many mysteries hidden in this concept as there are usually believed to be. These supposed mysteries have rendered the Calculus of the infinitely small quite suspect to many people. Those doubts that remain we shall thoroughly remove in the following pages where we shall explain the calculus.

But the attempt made thereafter to remove the doubts was not found satisfactory at all. The metaphysical question at the root was deeper than what Euler thought and could not be resolved by the kind of explanation he gave of it.

In 1758, John Landen, an English surveyor, remembered for his work in the theory of elliptic integrals, also attempted to give his understanding of what the infinitely small is. He gave it in his book

Discourse Concerning the Residual Analysis where he defined his derivative by the residue

$$\left[\frac{f(x_1) - f(x_0)}{x_1 - x_0}\right]_{x_0 = x_1}$$

expanding $f(x)$ in a power series in x.

This treatment did not satisfy many. But Lagrange, who was working on similar lines, gave a full explanation of his ideas on the metaphysical problem of infinitesimal calculus, basing his re-evaluation of the whole problem on Taylor expansion. In his *Theorie Des Functions Analytiques* published in 1799, he remarks

> The term function was used by the first analysts in order to denote in general the powers of a given quantity. Since then the meaning of this term has been extended to any quantity formed in any manner from any other quantity. Leibnitz and the Bernoulli were the first to use it in this general sense, which is nowadays the accepted one.

He thus brought another basic concept of calculus, namely, the term *function* in the field of discussion. This discussion, conducted with the object of arriving at the broadest acceptable meaning for the term, was also most warmly conducted between D'Alembert, Euler and Lagrange. Lagrange then goes on:

> Let us assign to the variable of a function some increment by adding to this variable an arbitrary quantity i. If the function is algebraic, we can expand it in terms of the powers of this quantity by using the familiar rules of algebra. The first term of the expansion will be the given function which will be called the primitive function : the following terms will be formed of various functions of the same variable multiplied by the successive powers of the arbitrary quantity. These new functions will depend on the primitive function from which they are derived and may be called the *derivative* functions. Generally speaking, whether the primitive function is

> algebraic or not, it can always be expanded in the same manner, and in this way it will give rise to the derived functions. The functions considered from this point of view, lead to an analysis superior to the ordinary one because of its generality and its numerous applications and we shall see in this work that the analysis that is commonly called transcendental or *infinitesimal* is, in fact, not different from that of the primitive and derivative functions : and that the differential and integral calculus is also, properly speaking, nothing but the calculus of these very same functions.

Lagrange elaborated all this in the next few pages and even illustrated his remarks by demonstrating at length the algebraic work which brings out the primitive and the derivatives of various orders. But, as compared with the way in which Newton and the earlier workers had defined a function and its derivative even with its suspect logic, appeared far too simple, short and direct for Lagrange's analysis to become easily acceptable.

4. The lead.

The lead came indirectly from Gauss. In 1812, he published his work on the *hypergeometric series*

$$1+\frac{\alpha\cdot\beta}{1\cdot\gamma}x+\frac{\alpha(\alpha+1)\beta(\beta+1)}{1\cdot 2\cdot\gamma(\gamma+1)}x^2+\cdot$$

To find its *region of convergence*, he applied the ratio-test to it and found that since

$$\left|\frac{a_{n+2}}{a_{n+1}}\right|=\left|\frac{(\alpha+n)(\beta+n)}{(n+1)(\gamma+n)}\right|\cdot|x|$$

whose limit as $n\to\infty$ is $|x|$, the series *converges absolutely* for $|x|<1$ and diverges for $|x|>1$.

Gauss did not stop at this. He tested its convergence at 1 and at -1, the extremities of its interval of convergence, for which, he had to make a *special investigation* and *devise* for it what is now called the *Gauss' test*, namely:

$$\text{If} \qquad \frac{a_n}{a_{n+1}} = \frac{n^p + \alpha_1 n^{p-1} + \cdots + \alpha_p}{n^p + \beta_1 n^{p-1} + \cdots + \beta_p}$$

where $p, \alpha_1, \ldots, \alpha_p, \beta_1, \ldots, \beta_p$ are independent of n, then

$$\sum a_n \text{ is convergent if } \alpha_1 - \beta_1 > 1 \text{ and}$$

$$\text{divergent if } \alpha_1 - \beta_1 \leq 1,$$

and showed that when $x = 1$, the hypergeometric series converges only when $\alpha + \beta - \gamma < 0$, and then absolutely; and further that when $x = -1$, it converges only when $\alpha + \beta - \gamma - 1 < 0$ and absolutely when $\alpha + \beta - \gamma < 0$.

The thoroughness with which Gauss had completed this work, in a way, gave an impressive lead to the workers in the field and indicated precisely how thorough and careful our work ought to be when it involves the *infinite processes*.

The next to follow was the noted analyst A. L. Cauchy who published, in 1821, his path-setting book *Cours D'Analyse*. It is in this book that we find for the first time a perfectly acceptable definition of *limit*, the all-important basic term of Analysis. Cauchy gave the definition in the following manner:

If, given any $\epsilon > 0$, there exists a $\delta > 0$, such that

$$|f(x) - L| < \epsilon \;\text{ when }\; 0 < |x - a| < \delta$$

then we say:

$$L \text{ is the limit of } f(x) \text{ as } x \to a.$$

Cauchy followed this by defining continuity, derivative, and definite integral of $f(x)$ in terms of the above definition of limit. Broadly speaking, Cauchy's definitions and explanations of the concepts involved were perfect and in this matter, Cauchy still rules the field. What good text-books carry today in the area of limits, continuity, differentiability, integrability and convergence of series and integrals is all Cauchy's work.

But Cauchy did not very seriously consider the problem of what a real number is. He took, more or less, an intuitive geometric view of real numbers and did not pay very much attention to the deeper distinctive properties of real numbers. This attitude of his was soon discovered to be too narrow and too superficial. In 1874, Weierstrass constructed the following ingenious function $f(x)$, where

$$f(x) = \sum_{n=0}^{\infty} b^n \cos(a^n \pi x)$$

where, a is an odd positive integer, $0 < b < 1$ and $ab > 1 + 3\pi/2$.

This function is continuous for all x but not differentiable at any x. This means that the curve it represents is continuous everywhere but *has no tangent at any point* – an observation quite contrary to our general intuitive geometrical understanding of curves. The example made one thing particularly imperative : that in Analysis one must completely avoid reliance on intuition; and one cannot, therefore, take real numbers to be more or less granted but must search the system thoroughly for deeper *arithmetical* properties on which the ultimate must rest.

Even earlier than this, in the 1850s, Riemann had demonstrated that the condition of continuity which Cauchy had prescribed for integrability of functions was too broad. He constructed the function defined on [0, 1] by

$$g(x) = \begin{cases} 0 & \text{if } x \text{ is irrational} \\ 1/q & \text{if } x = p/q \text{ in its lowest terms} \end{cases}$$

which he showed to be continuous at all irrational values of x and discontinuous at all rational values and yet *integrable* over the interval [0, 1].

Examples like these made it clear that Real Numbers hold more secrets than it appears on the surface and that Cauchy had not gone to the roots of them. This made it increasingly clear that a more systematic search was necessary to uncover these suspected properties and that the search must be very thoroughly planned, organized and conducted.

5. The Search.

The search for an unimpeachable basis for Analysis was undertaken principally by Weierstrass and simultaneously by two other contemporary mathematicians, Dedekind and Cantor. It was organized in two parts:

ONE: To clearly define a Real Number, and to give the system of real numbers an impeccable structure which will guarantee the elimination of every possibility of any internal contradiction at any future date.

TWO: To give rigorous definitions, based only on arithmetic, of terms such as a limit, continuity, differentiability, integrability and convergence involved in infinite and infinitesimal processes.

Very little, indeed, had remained undone in so far as the second part was concerned. Cauchy had done most of it : and quite satisfactorily. The few places where Cauchy's intuitive understanding of real numbers had peeped in had to be modified. And it would have been easy to do it once a clear picture of real numbers had emerged which would have happened after work on the first part would have been completed.

Weierstrass' programme had therefore concentrated on the first part. It involved two basic problems : both of them fundamental and both of them challenging. The problems were:

ONE: How does one define a real number? What is the criterion to determine whether a given symbol (presumably for a real number) is a real number? For example, how does one decide that $\sqrt{2}$ or π or e are real numbers?

TWO: On what basic foundation should the structure of Real Numbers be erected so that there is *no possibility* of any internal contradiction springing up in the system at any future date?

In so far as this second part is concerned, there was a model before them. Hamilton had shown in his 1933 paper on the algebra of couples, how complex numbers $a + bi$ could be built on number-couples (a, b) of real numbers, defining equality, addition and multiplication suitably, so that all properties of complex numbers could be derived by working with these number-couples. Since complex numbers thus depended entirely on real numbers, there would have been no possibility of any internal contradiction in the system of complex numbers if there was no such contradiction amongst real numbers.

But how could one guarantee this for Real Numbers unless their structure was based on a more fundamental system in respect of which one could guarantee the absence of any such contradiction?

Now, there was only one such system of numbers in respect of which it was possible to guarantee with some confidence, the complete absence of internal contradictions. This was the system of *natural numbers*. Mathematics had used natural numbers and worked with them for hundreds of years and had had no evidence of contradictions in the system. Therefore, it was natural that all those who felt it necessary to attend to this question, decided to revert back to the dream of the Pythagoreans to *base all number-systems on the system of natural numbers.*

6. Construction of I from N and Q from I.

From the system **N** of natural numbers, they first *constructed* the system **I** of *integers*. This is how they did it.

A. Construction of I from N

If $m, n, s, t, \ldots$ are natural numbers, ordered pairs (m, n) were *called integers* if they satisfied the following rules of arithmetic

1. Equality:

$$(m, n) = (s, t) \quad \text{if and only if} \quad m + t = n + s$$

2. Addition:

$$(m, n) + (s, t) = (m + s, n + t)$$

3. Multiplication:

$$(m, n) \times (s, t) = (ms + nt, mt + ns)$$

4. Order:
   Integers (m, n) in which $m > n$ were said to form a subset called the set of *positive* integers.

This is not the place to show that integers constructed as above are the integers with which we are familiar. Nor is there a need to do it. For mathematics, *these are the integers*, hence-forward. It can be shown that *these integers* have all the properties which we know the older familiar system has; and text-books show it. Here, our sole intention was to show that integers as now constructed have been based *only* on natural numbers and would be, on that account, free from internal contradictions.

B. Construction of Q from I

Let $m, n, s, t, \ldots$ be integers. Ordered pairs (m, n) in which the second numbers $n \neq 0$, were called *rational numbers* if they satisfied the following laws of arithmetic:

1. Equality:

$$(m, n) = (s, t) \text{ if and only if } mt = ns.$$

2. Addition:

$$(m, n) + (s, t) = (mt + ns, nt).$$

3. Multiplication:

$$(m, n) \times (s, t) = (ms, nt).$$

and

4. Order:
   The subset of ordered pairs (m, n) in which $mn > 0$ were said to form a subset of *positive rational numbers.*

Remarks similar to those made above about integers, after their construction from natural numbers, equally well apply to the Rational Numbers constructed from integers. The question whether these are the same as the rational numbers, with which we are familiar, does not arise. For mathematics hence-forth, *these are the rational numbers.*

Let us only make two minor notes here, about the zero and the unity of the system. The rational number (m, n) in which $m = 0$ is the rational number, *zero.* And the rational number (x, x) is the rational number, *unity.* For convenience, the two are usually written $(0, 1)$ and $(1, 1)$ and finally as 0 and 1.

7. The search continued.

The real problem however had still remained unresolved. The whole effort was to find an unimpeachable foundation for **R**, the system of Real Numbers. And that task was not still done. The task specifically was to base the structure of **R** on either the set of rational numbers, or the set of integers or on the set of natural numbers, each one of which was now known to be free of internal contradictions. The task was difficult. But it was done. And surprisingly, it was done almost simultaneously, in two distinct ways. One method was developed by Cantor and the other by Dedekind. Cantor's was based on properties of Cauchy sequences of rational numbers. Dedekind's was based on his novel idea of *cuts* of rational numbers.

A. Cantor's Theory.

Cantor's construction of the Real Numbers through rational numbers runs broadly as follows. He first defines $\{x_n\}$, a Cauchy Sequence of rational numbers $x_1, x_2, x_3, \ldots$ as one for which

> Corresponding to every positive rational number ϵ, there exists a natural number N such that
>
> if m, n are natural numbers greater than N, then
>
> $$|x_m - x_n| < \epsilon.$$

For example, $\{\frac{n+1}{n}\}$ is a Cauchy sequence; because, corresponding to a positive rational number ϵ, one can, in this case, choose N

to be an integer greater than $\frac{1}{\epsilon} + 1$ and verify that

$$\begin{aligned} |x_m - x_n| &= |\frac{m+1}{m} - \frac{n+1}{n}| \\ &= |\frac{n-m}{mn}| \\ &= |\frac{1}{m} - \frac{1}{n}| \\ &< \frac{1}{m} \\ &< \frac{1}{N} \qquad \text{since} \quad N < m \\ &< \frac{1}{1/\epsilon} \qquad \text{since} \quad \frac{1}{\epsilon} < N \\ &= \epsilon \end{aligned}$$

As another example, consider $\{\frac{n+5}{n}\}$, which also can be similarly proved to be a Cauchy sequence by choosing N greater than $5/\epsilon + 1$.

As a second step, Cantor defined *equality* of two Cauchy sequences by prescribing that two Cauchy sequences $\{x_n\}$ and $\{y_n\}$ are equal if corresponding to any positive rational number ϵ, there exists an integer N such that

$$|x_n - y_n| < \epsilon \text{ for every } n > N.$$

Consider the two Cauchy sequences $\{\frac{n+1}{n}\}$ and $\{\frac{n+5}{n}\}$. They are equal, since, corresponding to a positive rational number ϵ, one may choose N to be greater than $4/\epsilon + 1$ and notice that

$$\left|\frac{n+1}{n} - \frac{n+5}{n}\right| < \epsilon \quad \text{for all } n > N.$$

In fact there are an infinite number of Cauchy sequences equal to any given Cauchy sequence. This infinite set of Cauchy sequences equal to $\{x_n\}$ is called an *equivalence class* of Cauchy sequences, the particular sequence $\{x_n\}$ being one representative member of this class. Such different equivalence classes would now be denoted by $x, y, u, \ldots$.

Cantor next considered sequences

$$\{s_n\} \text{ and } \{p_n\}$$

where $$s_n = x_n + y_n \text{ and } p_n = x_n y_n$$
and showed that $\{s_n\}$ and $\{p_n\}$ are Cauchy sequences. He called them the *sum* and the *product* respectively of the sequences $\{x_n\}$ and $\{y_n\}$ and wrote

$$s = x + y \text{ and } p = x \times y.$$

It is easy to see that these operations of *addition* and *multiplication* introduced in the set of Cauchy sequences are both *commutative* and *associative.*

Cantor next introduced two *special* Cauchy sequences. The first one which he defined as a Cauchy sequence

$$0, 0, 0, \ldots$$

each term of which is 0, was called *zero* and denoted by 0. The second special Cauchy sequence

$$1, 1, 1, \ldots$$

each term of which is 1, was called *unity* and denoted by 1. With zero and unity, one obtains the results

$$x + 0 = x$$

and

$$x \times 1 = x$$

easily verifiable with the help of the preceding definitions. These results imply that 0 and 1 are the *additive* and *multiplicative identities.* The Cauchy sequence $-x = \{-x_n\}$ is the *additive inverse* of $x = \{x_n\}$.

Suppose that x is such a Cauchy sequence that there exists a positive rational number ϵ and an integer N such that

$$|x_n| \geq \epsilon \quad \text{for every } n > N;$$

then, x is called a non-zero Cauchy sequence. And, it can be proved that for each such non-zero sequence x there exists a non-zero Cauchy sequence y such that

$$x \times y = 1$$

so that, every non-zero Cauchy sequence has a *multiplicative inverse.*

Thus it was shown that the set of Cauchy sequences constitute a *field.*

We next define a subset of Cauchy sequences which could be called a *subset of positive Cauchy* sequences. A Cauchy sequence x is said to be a *positive* Cauchy sequence if there is a positive rational number ϵ and an integer N such that

$$x_n > \epsilon \text{ for every } n > N.$$

All such positive Cauchy sequences together form the subset of positive Cauchy sequences; so that, now the field of Cauchy sequences becomes an *ordered field.*

The Cantor theory developed till this stage is sufficient to define a Real Number, the definition being:

> An *equivalence class of Cauchy sequences* of rational numbers is called a *Real Number.*

Defined as above, a number of properties of Real Numbers can now be established. The most basic property is contained in the following theorem:

> A non-empty set of Real Numbers, which has an upper bound has a *least upper bound.*

It was already noticed above that with properties of addition, multiplication, zero, unity, additive and multiplicative identities and inverses and the subset of positive elements, equivalence classes of Cauchy sequences of rational numbers constitute an ordered field. With the property that every set that is bounded above has a least upper bound, the above ordered field becomes a *complete ordered field.* This field, *by definition,* is **R** , the field of Real Numbers. Analysis is developed in this field. Concepts of limit, continuity, differentiability, integrability, convergence and similar other fundamental concepts of Analysis are developed in **R** constructed as above.

Thus, Cantor's construction of Real Numbers *based entirely on rational numbers* fulfills the original objective of Weierstrass' programme. In this theory,

> Every Cauchy sequence of rational numbers is a Real Number, and every Real Number is a Cauchy sequence of rational numbers.

As an illustration, consider the sequence s_n where

$$s_n = 1 - \frac{1}{3} + \frac{1}{5} - + \cdots + \frac{(-1)^n}{2n+1}.$$

We shall first show that it is a Cauchy sequence. Here,

$$\begin{aligned} |s_m - s_n| &= \left| \frac{(-1)^{n+1}}{2n+1} + \frac{(-1)^{n+2}}{2n+5} + \cdots + \frac{(-1)^m}{2m+1} \right| \\ &= \left| \frac{1}{2n+1} - \frac{1}{2n+5} + - \cdots + \frac{(-1)^m}{2m+1} \right| \end{aligned}$$

The sum between the absolute-value signs can be written in the form

$$\left(\frac{1}{2n+3} - \frac{1}{2n+5}\right) + \left(\frac{1}{2n+7} - \frac{1}{2n+9}\right) + \cdots$$

$$\cdots + \begin{cases} \left(\dfrac{1}{2m-1} - \dfrac{1}{2m+1}\right) & \text{when } m-n \text{ is odd} \\ \dfrac{1}{2m+1} & \text{when } m-n \text{ is even} \end{cases}$$

Since $\{1/(2n+3)\}$ is decreasing, this shows that the sum is ≥ 0; and therefore the absolute-value signs may be removed. If this sum is written in the form

$$\frac{1}{2n+3} - \left(\frac{1}{2n+5} - \frac{1}{2n+7}\right) - \cdots$$

$$\cdots - \begin{cases} \left(\dfrac{1}{2m-1} - \dfrac{1}{2m+1}\right) & \text{when } m-n \text{ is odd} \\ \dfrac{1}{2m+1} & \text{when } m-n \text{ is even} \end{cases}$$

then this shows further that the sum is $\leq 1/(2n+3)$.

Since $1/(2n+3)$ decreases, it is less than any positive rational number ϵ for all n larger than $(N-3)/2$ where N is chosen larger than $1/\epsilon$.

Thus the given sequence is a Cauchy sequence.

Therefore, by definition, it is a Real Number. We designate this Real Number by the symbol

$$\frac{\pi}{4}.$$

The "number" π with which mathematics has so much to do and which was for hundreds of years taken for granted as a real number, is thus *shown* to be a *real number.*

Consider the illustration of another sequence s_n where

$$s_n = \frac{5n^2+3}{8n^2}$$

which, on simplification, gives

$$s_n = \frac{5}{8} + \frac{3}{8n^2},$$

a general term of a sum of a Cauchy sequence $\{5/8\}$, all of whose terms are $5/8$ and another Cauchy sequence $\{\frac{3}{8n^2}\}$ equal to sequence 0. Thus the sequence is the sum of two real numbers which could be in a natural way designated as $5/8$ and 0. The given sequence is therefore the real number $5/8+0$, i.e., $5/8$.

B. Dedekind Theory of Cuts.

Dedekind came upon the same problem in a very different context. In the preface to his pamphlet *Continuity and Irrational Numbers*, published in March 1872, he writes:

> My attention was first directed towards the considerations which form the subject of this pamphlet in the autumn of 1858. As professor in the Polytechnic School in Zurich, I found myself for the first time obliged to lecture upon the elements of differential calculus and felt more keenly than ever before the lack of a really scientific foundation for arithmetic. In discussing the notion of the approach of a variable magnitude to a fixed limiting value, and especially in proving the theorem that every magnitude which grows continually, but not beyond all limits, must certainly approach a limiting value I had recourse to geometric evidences. Even now such resort to geometric intuition in a first presentation of the differential calculus, I regard as exceedingly useful, from the didactic standpoint, and indeed indispensable, if one does not wish to loose too much time. But that this form of introduction into the differential calculus can make no claim to being scientific, no one will deny. For myself this feeling of dissatisfaction was so overpowering that I made the fixed resolve to keep meditating on the question till I should find a purely arithmetic and perfectly rigorous foundation for the principles of infinitesimal analysis. The statement is so frequently made that the differential calculus deals with continuous magnitude, and yet an explanation of this continuity is nowhere given; even the most rigorous expositions of the differential calculus do not base their proofs upon continuity but, with more or less consciousness of the fact, they either appeal to geometric notions or those suggested by geometry, or depend upon theorems which are never established in a purely arithmetic manner. Among these, for example, belongs the above mentioned theorem, and a more careful investigation

> convinced me that this theorem, or any other equivalent to it, can be regarded in some way as a sufficient basis for infinitesimal analysis. It then only remained to discover its true origin in the elements of arithmetic and thus at the same time to secure a real distinction of the essence of continuity. I succeeded on November 24, 1858.

Conceptually Dedekind's theory is much simpler. Let us begin with the directed number line XY and locate on it the integer 0.

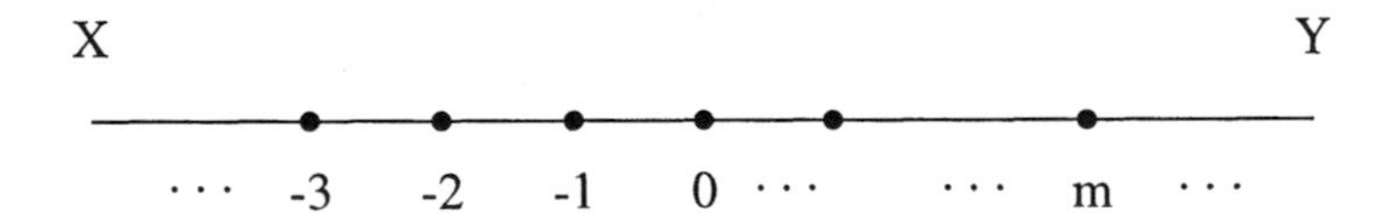

Choosing a convenient scale, let us locate numbers 1, 2, 3, ..., m, ... on line XY to the right of 0 and integers ..., -3, -2, -1, to the left of 0 proceeding from left to right.

To locate a positive rational number m/n on it, we draw a line

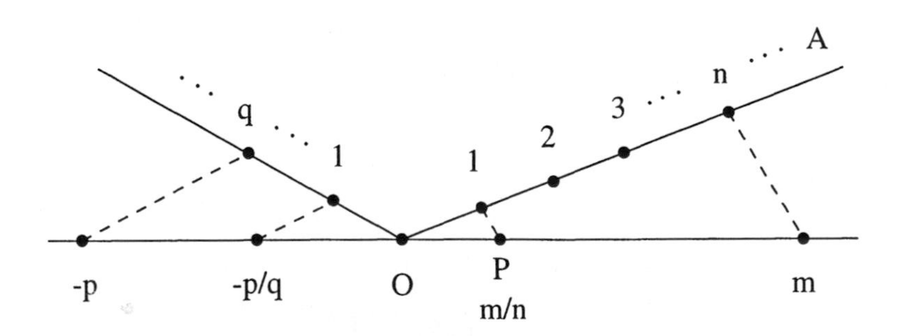

through 0, and on this line choose points at distances 1, 2, 3, ..., n from point 0. If we join point n to point m and draw through point 1 a line parallel to nm to meet XY in point P, the point P is a representation on the number line of the required rational number m/n. If it is a negative rational number $-p/q$, a construction similar to the one for m/n shall have to be done in the left half of the figure.

Thus *all* rational numbers can be shown on a directed number-line. Let XY below be such a line. Take a cut at number 3. This

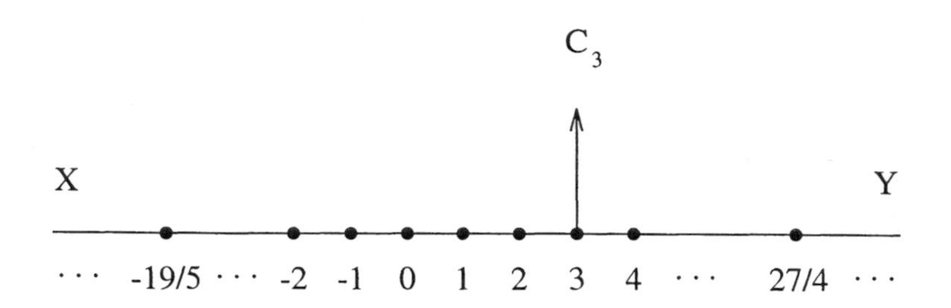

cut shall put all rational numbers *less* than 3 to the left of 3. This class of rational numbers is called the *left* class of the cut. The cut also puts all rational numbers greater than 3 to the right of 3. This class of rational numbers is called the *right* class of the cut. As a convention, we put rational number 3 in the right class. The cut itself is named cut C_3 and is called a Dedekind Cut.

This is a broad picture of a Dedekind Cut, so described with reference to a number-line. This makes it easy to understand and remember the essential characters of the cut; namely, the Left and the Right classes and the fact that elements of one class are, one and all, less than the members of the other class. We now give the definition in more precise terms.

> A division (L, R) of the set **Q** of all rational numbers is called a *Dedekind Cut* if it has the following three properties:

1. Neither L nor R is an empty set.

2. if a is a rational number in L, and b is a rational number less than a, then b is in L. Similarly, if c is a rational number in R and d is a rational number larger than c, then d is in R. and

3. If a is a rational number in L, then there exists a rational

number g which is larger than a and which *lies in* L. That is, the class L has *no greatest member.*

Let us check these properties in respect of the cut introduced above as C_3.

1. Rational number 1 is in L, so that L is not empty. Rational number 7 is in R so that R is not empty. Thus property 1 is satisfied.

2. Suppose a rational number a is in L. Since it is in L, $a < 3$. If b is less than a, then b is less than 3; and therefore b lies in L. Similarly, if c is in R, then $c \geq 3$. If $d > c$, then d is larger than 3 and would therefore be in R. Thus property 2 is satisfied.

3. Suppose a is a rational number in L, so that $a < 3$. Now between two distinct rational numbers, there lie other rational numbers. Since a and 3 are distinct, there is a rational number g between them; so that we have $a < g < 3$. Since $g < 3$, g is in L. And property 3 is satisfied.

Since C_3 satisfies all the three prescribed properties, C_3 is a proper Dedekind Cut. In the case of this cut, it *happens* incidentally that the R of C_3 has a *least member*, namely, the rational number 3.

However, it could happen, and indeed, in many cases, it happens that the Right class R of a Dedekind Cut does not have a Least Member. Some cuts may have this property; some may not. And it was Dedekind's remarkable insight that he could detect this fact and could bring it to the notice of all his contemporaries who could see its far-reaching consequences.

Consider the following illustration of this fact.

Let there be a division of the set of all rational numbers **Q**.

Let L be the set of all negative rational numbers, the rational number 0 and those positive rational numbers whose squares are less than 2. Let R consist of those positive rational numbers whose squares are greater than 2.

Now, since **Q** consists of rational numbers, which are either negative or 0, and positive rational numbers whose squares are either less than 2 or greater than 2, *all* rational numbers participate in the partition proposed above. The first two properties of a Cut are also easily seen to be true of this partition. The problem is only of the third property. Does L have or does L not have a *greatest* member. To be a Cut, L must have no greatest member. We show now that this condition is, indeed, satisfied.

If L has a greatest, it would be some positive member of L. Let us check this for positive members of L. Let a be a positive rational number in L. Now consider a rational number b, such that

$$b = a + \frac{2 - a^2}{3 + a}.$$

Obviously, $b > a > 0$. We shall now show that b is in L. Actually,

$$b = a + \frac{2 - a^2}{3 + a} = \frac{3a + 2}{3 + a}$$

and

$$\begin{aligned} 2 - b^2 \qquad &= 2 - \frac{9a^2 + 12a + 4}{9 + 6a + a^2} \\ &= \frac{14 - 7a^2}{(3 + a)^2} \\ &= \frac{7(2 - a^2)}{(3 + a)^2} \quad > 0 \end{aligned}$$

Thus $b^2 < 2$ and therefore b is in L. Thus if a is in L, a larger number b is also in L and L has no greatest member; so that the third property of a cut is also satisfied. The cut, thus, is a Dedekind Cut.

Interestingly, and unlike the R of C_3, the class R of this cut *has no least.*

To see this, consider u to be a member of the right class R; so that $u > 0$ and $u^2 > 2$. That is,

$$\frac{2}{u} < u$$

and

$$\left(\frac{u + 2/u}{2}\right) < \frac{u + u}{2} = u$$

and since,

$$\left(\frac{u}{2}+\frac{1}{u}\right)^2=\left(\frac{u}{2}-\frac{1}{u}\right)^2+2>2,$$

the rational number $(\frac{u}{2}+\frac{1}{u})$ is less than u and has its square greater than 2. This means that whatever rational number u belongs to R, a *smaller* rational number $\frac{u}{2}+\frac{1}{u}$ also belongs to R. So that R has *no least.*

Thus Dedekind demonstrated that:

> There are two types of cuts. In one, R has a least member. In the other, R has no least member.
>
> Cuts of the former type were called *rational real numbers.*
>
> Cuts of the latter type were called *irrational real numbers.*

Of the two cuts shown above, the first, C_3, is identified with the rational real number 3; the second with "Real Number $\sqrt{2}$".

This was the main contribution of Dedekind. What remained still to be done was the formal work. As with the set of Cauchy sequences of rational numbers, we now take up K, the set of *all cuts.* We define a relation of equality in K and the operations of *addition* and *multiplication.* We next show the existence of additive and multiplicative identities and inverses. We then show that a subset of K can be defined as a subset of *positive cuts.* And finally we show that that this set K now is an *ordered field.*

This is followed by the proof of the basic fundamental Theorem that every subset of K which is bounded above has a *least upper bound.* This theorem makes the set of Dedekind Cuts a *complete ordered field,* which then is identified with the field of Real Numbers.

It is remarkable that this most intriguing problem of *constructing real numbers* from Rational Numbers, was solved in *two* distinct and equally admirable ways.

The movement to find an unimpeachable basis for Analysis, thus came to a successful end. The search took almost a hundred years and ended in 1872 with the paper *Continuity and Irrational Numbers* of Dedekind. The final success was, indeed, a great achievement, an achievement of such brilliant mathematicians as

D'Alembert, Gauss, Cauchy, Weierstrass, Cantor and Dedekind.

The whole work and the way it proceeded was called by Felix Klein in 1895 the

Arithmetization of Analysis.

Exercises

1. (m, n) and (s, t) are positive integers. Show that

   (a) $(m, n) + (s, t)$ is a positive integer.

   (b) $(m, n) \times (s, t)$ is a positive integer.

2. (a) Let (a, b) be a positive integer and $(m, n) > (s, t)$. Show that

   $$(a, b) \times (m, n) > (a, b) \times (s, t)$$

   (b) Let (a, b) be a negative integer and $(m, n) > (s, t)$. Show that

   $$(a, b) \times (m, n) < (a, b) \times (s, t)$$

3. If (m, n) is a positive integer and (s, t), a negative integer, show that $(m, n) \times (s, t)$ is a negative integer.

4. If (m, n) and (s, t) are both negative integers, show that $(m, n) \times (s, t)$ is a positive integer.

5. Show that multiplication of rational numbers is

   (a) commutative
   (b) associative
   and (c) distributive over addition.

6. If rational numbers (m, n) and (s, t) are both positive and less than unity, show that their product is also less than unity.

7. If rational number (m, n) is positive and rational number (s, t) is negative, show that $(m, n) \times (s, t)$ is negative.

8. Show that the set **Q** of rational numbers constructed as above is *closed* under addition as well as multiplication.

9. Show that for all natural numbers m, n, $n \neq 0$, the square of the rational number (m, n) is non-negative.

10. (a) Show that the sequence, s_n, where

$$s_n = 1 + \frac{1}{1!} + \frac{1}{2!} + \frac{1}{3!} + \cdots + \frac{1}{n!}$$

is a Cauchy sequence and hence a real number.

(b) If this real number is A, show that $2 < A < 3$.

11. Consider the following division of rational numbers:

(i) Put in L all negative rational numbers, the rational number 0 and all those positive rational numbers whose squares are less than 5.

(ii) Put in R all the remaining positive rational numbers.

(a) Show that this is a Dedekind Cut.

(b) Show further that this Cut represents an irrational real number.

12. Let p be a prime number. Consider the following division of all rational numbers:

(i) Put in class L all negative rational numbers, the rational number 0 and all those positive rational numbers whose squares are less than p.

(ii) Put in R all the remaining positive rational numbers.

(a) Show that this is a Dedekind Cut.

(b) Show also that this Cut is an irrational real number.

Further Reading

1. *Set Theory and Logic*, Robert R. Stall, W. H. Freeman and Company, San Francisco and London, 1963.

2. *Introduction to Modern Algebra*, N. H. McCoy, Allyn and Bacon, Boston, 1965.

3. *Continuity and Irrational Numbers*, Richard Dedekind, Tr. W. W. Beman, Open Court Publishing Company, Chicago, 1909.

CHAPTER EIGHT

BEGINNINGS OF ALGEBRA

1. Al-Khowarizmi and Emmy Noether.

To his book on the history of Algebra, B. L. Van der Waerden has given the title

A HISTORY OF ALGEBRA
From Al-Khowarizmi to Emmy Noether.

In fact, neither Al-Khowarizmi was the first nor Emmy Noether the last of the great algebraists. Both however, in their own ways, were remarkable personalities. It was on account of the former that Algebra came to Europe : it was on account of the latter that Algebra acquired its sharpest tone. But apart from this one parallel in their lives, there was none else. Al-Khowarizmi flourished as an esteemed friend of Caliph Al-Mamun of Baghdad and was an extremely respected member of the *Dar Al-Hikman* (the *House of Wisdom*), the Royal Academy of Baghdad. Emmy Noether, on the other hand, though a reputed algebraists of Gottingen who had the distinction of having such algebraists as B. L. Van der Waerden and P. S. Alexandroff as her students, had to leave Gottingen because of fear of royal displeasure. Eminent scientists such as Einstein and Courant had to manage her shift to America. In America, she taught in a college at Bryn Mawr near Princeton but met with an unfortunate early death. When she died in 1935 at the age of fifty-three, Einstein wrote a letter about her in New-York Times to which the newspaper gave a place of pride on its front page, an honour considered extremely rare.

2. Al-Khowarizmi's book.

Algebra was cultivated in India much earlier than this and to a much higher level. But at that time it could not have easily reached the shores of Europe and entered the main stream of mathematics. Firstly because Indian mathematicians wrote in *Sanskrit* which had by then ceased to be the spoken language of the people and had become the guarded preserve of the sophisticated few. Secondly, they wrote their mathematics in the form of verses, for which reason it had to be brief and cryptic and consequently all the more incomprehensible. And lastly there was no regular concourse, either social or commercial, between India and the countries of the southern Mediterranean shores of Europe.

Al-Khowarizmi was conversant with Indian mathematics. He had studied the Indian numerals and the algorithms of Indian arithmetic and even written a short book on the same. At the request of Al-Mamun and with the help of an Indian astronomer who then was a royal guest at Baghdad, he had rendered into Arabic a Sanskrit text *Sindhind* on astronomy.

He was conversant, as most say he was, with the *Bijaganita* (Algebra) of Brahmagupta and Bhaskaracharya.

Whether he was or he was not, the book

Al-Jabr wul Al-Muqabala

which he wrote in Arabic is entirely his own. It is remarkably refreshing and leisurely. Written in simple Arabic prose, Al-Khowarizm was at no stage in a hurry to slur over any single important point. He takes his own time, explains every intriguing point at length, illustrates the point with examples wherever necessary and makes the book altogether readable. Written in Arabic which was the spoken language of the people of the region and with which the European traders and merchants from the Mediterranean shores of Europe were quite conversant, the book could come to the easy notice of the Europeans. And its contents could naturally find a place in the main stream of European mathematics.

In the prefatory part of the book, he pays his deep respects to his master the *Imam Al-Mamun* and explains the objectives which the Imam had placed before him in order that the book may be useful to all. In his words,

> The fondness for science, by which God has distinguished the *Imam Al-Mamun*, ..., the affability and condescen-

sion which he shows to the learned, the promptitude with which he protects and supports them in the elucidation of obscurities and in the removal of difficulties — these qualities of his head and heart, have encouraged me to compose a short work on calculating by the rules of *completion* (Al-jabr) and *reduction* (Al-Muqabala), confining it to what is easiest and most useful in arithmetic, such as men constantly require in cases of inheritance, legacies, partitions, law-suits and trade, and in all their dealings with one another, or where the measuring of lands, the digging of canals, geometrical computations and other objects of various sorts and kinds are concerned ...

3. Al-jabr and Al-muqabala.

The title

Al-Jabr wul Al-Muqabala

of Al-Khowarizmi's book sounds strange but is distinctive. Actually the words al-jabr and al-muqabala are ordinary words from the Arabic language; but since the time of the book of Al-Khowarizmi, these words have acquired special significance. In fact, the word

al-jabr

is what has earned the name

algebra

for the part of mathematics which now is called Algebra.

The usual meaning with which the word *jabr* is used in Arabic texts on mathematics is:

> *adding equal terms to both sides of an equation in order to eliminate negative terms*

and the word *muqabala* is used with the meaning

> *reduction of positive terms by subtracting equal amounts from both sides of an equation.*

Thus, the treatise of Al-Khowarizmi is a collection of methods of solving problems by using these two devices of *completion* (al-jabr) and *reduction* (al-muqabala).

4. Six basic formulae.

The first results which Al-Khowarizmi establishes in the book are the

six standard forms

to which given equations can be reduced and the corresponding six formulae for their solutions. He does it through six problems as below.

Consider the first problem:

> I have divided ten into two portions; I have multiplied the one of the two portions by the other; after this I have multiplied the one of the two by itself; and the product of the multiplication by itself is four times as much as that of one of the portions by the other. Find the two parts concerned.

Now, Al-Khowarizmi does not use symbols like x and x^2 or even 10 and 40 but gives the whole computation in words as below:

> Computation: Suppose one of the portions to be thing, and the other ten minus thing : you multiply thing by ten minus thing; it is ten things minus a square. Then multiply it by four, because the problem states four times as much. The result will be four times the product of one of the parts multiplied by the other. This is forty things minus four squares. After this you multiply thing by thing, that is to say, one of the portions by itself. This is a square, which is equal to forty things minus four squares. Reduce it now by the four squares, and add them to the one square. Then the equation is: forty things are equal to five squares; and one square will be equal to eight roots, that is, sixty four; the root of this is eight, and this is one of the two portions, namely,

that which is to be multiplied by itself. The remainder from the ten is two, and that is the other portion. Thus the question leads you to one of the six cases, namely, that of "squares equal to roots". Remark this.

If notation was used, as we do now, the computation would have been as under : much shorter and much more compact.

He has divided 10 into two portions; we would denote these two portions by

$$x \text{ and } 10 - x$$

x denoting what Al-Khowarizmi has referred to as *thing*. Since, now he has multiplied thing by ten minus thing, that is, since x is multiplied by $10-x$, the product would be $x \times (10 - x)$, that is,

$$10x - x^2$$

the same as Al-Khowarizmi has referred to as ten things minus a square. Then he multiplies this by 4 to get the product as

$$40x - 4x^2.$$

After this, he multiplies thing by thing; that is x by x and gets

$$x^2$$

as the result. Since this was equal to forty things minus four squares, he gets an *equation*

$$x^2 = 40x - 4x^2.$$

At this stage he uses the method of *Al-jabr*, adds $4x^2$ to both sides and reduces the equation to

$$5x^2 = 40x$$

a form, to which Al-Khowarizmi refers as "squares equal to roots". He solves it as below; from $5x^2 = 40x$ he gets $x^2 = 8x$ and from this last equation he gets the answer

$$x = 8.$$

Even this solution, which we have given above, is unnecessarily lengthy. It has been so shown to bring out, as emphatically as one can, the important observation that

Good notation is a tremendous help to good mathematics.

But notation came to Algebra much later; after another few hundred years. The first to be systematically employed was introduced by Vieta.

Also note the phrase "Remark this" used in the last part of Al-Khowarizmi's solution, where he says:

> Thus the question leads you to one of the *six* cases, this first one typified in his language as
>
> *squares equal to roots.*

In the first forty-seven pages of the book Al-Khowarizmi has made a list of six problems such as the first one given above. Computation of their solutions are given. These six problems are very carefully framed and they lead to six distinct standard types, such as the type "squares equal to roots" to which the first problem led. Let us go over his six problems and demonstrate how each one of them leads to a different type and how together they represent all possible types. We shall also thereby get acquainted with the kind of language and technical terms which they were using at that time.

The *second* problem is:

> I have divided ten into two portions; I have multiplied each of the parts by itself, and afterwards ten by itself the product of ten by itself is equal to one of the two

> parts multiplied by itself, and afterwards by two and seven-ninth.

Using our notation, this problem firstly becomes

$$10^2 = x^2 \times 2\frac{7}{9}$$

and finally reduces to

$$36 = x^2$$

that is, to the second type

squares equal to numbers

in Al-Khowarizmi's terminology.
His third problem is:

> I have divided ten into two parts. I have afterwards divided the one by the other, and the quotient was four.

In our notation, the problem leads to

$$\frac{10 - x}{x} = 4$$

which finally gives

$$10 = 5x$$

which is the third type

roots equal to numbers

in Al-Khowarizmi's language.
His fourth problem is:

> I have multiplied one-third of thing and one dirhem by one-fourth of thing and one dirhem, and the product was twenty.

In our notation, the above problem is to find x when

$$(\frac{1}{3}x + 1)(\frac{1}{4}x + 1) = 20$$

which finally reduces to

$$x^2 + 7x = 228$$

which is Al-Khowarizmi's fourth type

squares and roots equal to numbers.

Al-Khowarizmi has listed two more typical problems. His *fifth* problem is:

> I have divided ten into two parts; I have multiplied each of them by itself, and when I had added the two products together, the sum was fifty-eight dirhems.

Using x and x^2 and other notation, this problem amounts to finding x when

$$x^2 + (10 - x)^2 = 58$$

which, when simplified, leads to equation

$$x^2 + 21 = 10x,$$

the fifth form

squares and numbers equal to roots

in the terminology of Al-Khowarizmi.

Finally he comes to his *sixth* and last type, to illustrate which he takes the problem:

> I have multiplied one-third of a root by one-fourth of a root and the product is equal to the root and twenty-four dirhems.

In terms of our notation, Al-Khowarizmi's sixth problem becomes one of finding x when

$$\frac{x}{3} \times \frac{x}{4} = x + 24$$

which, when simplified, leads to finding x when

$$x^2 = 12x + 288$$

to which last form of his, Al-Khowarizmi gives the name

roots and numbers equal to squares.

5. Standard Formulae.

To start with, thus, Al-Khowarizmi enumerates all possible standard forms to which problems get reduced. At present we have only two forms, namely, the linear and the quadratic; Al-Khowarizmi makes them six, because, he admits *only positive numbers* as coefficients of the squares and of the roots and as the numbers. By this means he has made his formulae less intriguing and consequently easier to remember. Pedagogically, this was a wise step in those early days of mathematics.

Actually, the idea behind the enumeration of these six forms at the very beginning of the treatise is to make it clear to those who would be using these methods that their efforts should be directed towards reducing their problems to one of these forms and then just *write the answers* according to the formulae. We shall give these formulae in our modern notation. Thus,

1. The first formula is:

   If $x^2 = ax$,
   the answer is: $x = a$.

2. The second formula is:

   If $x^2 = a$,
   the answer is: $x = \sqrt{a}$.

3. The third formula is:

   If $ax = b$,
   the answer is: $x = b/a$.

4. The fourth formula is:

   If $x^2 + ax = b$,
   the answer is: $x = \sqrt{a^2/4 + b} - (a/2)$.

5. The fifth formula is:

   If $x^2 + a = bx$,
   the answer is: $x = (b/2) \pm \sqrt{b^2/4 - a}$.

6. The sixth formula is:

If $x^2 = ax + b$

the answer is: $x = (a/2) + \sqrt{a^2/4 + b}.$

6. Illustrations of the solution of problems.

Al-Khowarizmi has followed this with a number of illustrations. One such is:

> I have divided ten into two parts; I have multiplied one of the two by the other, and have then divided the product by what the difference of the two parts was before their multiplication, and the result of this division is five and one-fourth.

When expressed in modern notation, the problem takes the form

$$\frac{x(10-x)}{(10-x)-x} = 5\frac{1}{4}$$

that is,

$$\frac{10x - x^2}{10 - 2x} = 5\frac{1}{4}$$

that is,

$$10x - x^2 = 52\frac{1}{2} - 10\frac{1}{2}x$$

so that, ultimately we get

$$20\frac{1}{2}x = x^2 + 52\frac{1}{2},$$

the fifth standard form of Al-Khowarizmi. The answer, therefore, is

$$x = 10\frac{1}{4} \pm \sqrt{\left(\frac{41}{4}\right)^2 - 52\frac{1}{2}}$$

$$= 10\frac{1}{4} \pm \frac{29}{4}$$

Hence

$$x = 10\frac{1}{4} - \frac{29}{4}.$$

The other root, namely $10\frac{1}{4} + \frac{29}{4}$, is larger than 10 and has to be rejected. Al-Khowarizmi does not specifically state this; but just shows that the root is

$$\frac{41}{4} - \frac{29}{4} = 3.$$

7. Mensuration Problems.

In the second part of the book, Al-Khowarizmi considers problems leading to measurement of land. The figures which he has considered include the triangle, the rectangle, the circle and their combinations. He has established the formulae for the areas of a right-angled triangle, an equilateral triangle and a triangle in general. One of his problems is:

> Find the area of a triangle whose three sides are thirteen, fifteen and fourteen.

His solution is:

> Let the triangle be the one in the figure If the height be

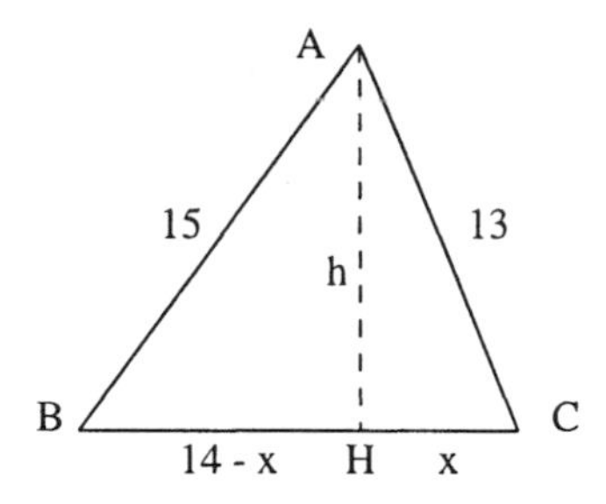

> AH, let AH $= h$, CH $= x$, so that BH is $14 - x$. Applying Pythagoras theorem to triangles AHC and

ABH, one gets

$$13^2 - x^2 = h^2 = 15^2 - (14 - x)^2$$

leading to

$$x = 5 \quad \text{and} \quad h = 12$$

so that the area of the triangle is

$$\frac{1}{2} \times 14 \times 12$$

that is 84.

A second problem, involving volumes, is

> A pyramid has a base of 4 yards; its height is 10 yards, and the dimensions of its upper extremity two yards by two yards. Find its bulk.

His solution is:

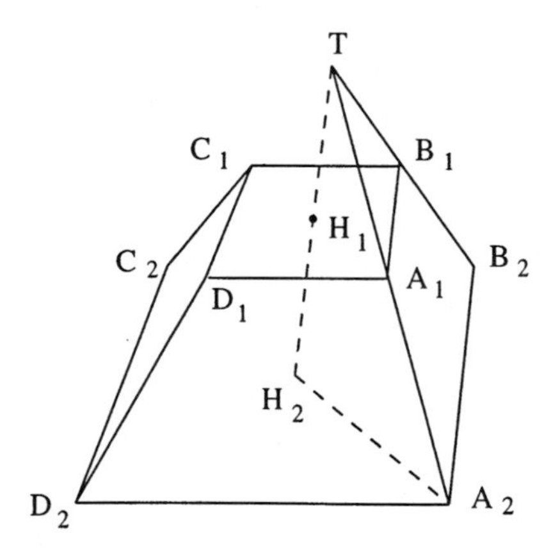

> We know that every pyramid is decreasing towards its top, and that one-third of the area of its basis multiplied by the height gives its bulk.

The present pyramid has no top. We must therefore seek to ascertain what is wanting in its height to complete the top. We observe, that the proportion of the entire height to the ten, which we have now before us, is equal to the proportion of four to two. Therefore, the whole height of the pyramid is 20 since $H_1H_2 = 10$ and $TH_1 = 10$ and

$$\frac{TH_2}{10} = \frac{TH_2}{H_1H_2} = \frac{TA_2}{TA_1} = \frac{A_2B_2}{A_1B_1} = \frac{4}{2} = 2$$

Now, the volume of the given pyramid

$= \text{Vol. of Pyr. } TC_2D_2A_2B_2 - \text{Vol. of Pyr. } TC_1D_1A_1B_1$

$= (1/3) \times 4 \times 4 \times 20 - (1/3) \times 2 \times 2 \times 10$

$= (1/3)(320 - 40)$

$= 93\frac{1}{3}.$

8. Problems of legacies.

Al-Khowarizmi has devoted the next section of the book to illustrate how algebra could help compute amounts marked out in a Will as legacies to the survivors. One such problem solved by Al-Khowarizmi is a model for those who have to ultimately deal with such problems in or outside a court of law.

The problem is:

> A man dies, and leaves four sons, and bequeathes to some person as much as the share of one of his sons; and to another, one-fourth of what remains after the deduction of the above share from one-third. Calculate the various bequests.

Al-Khowarizmi's solution in his language:

> Take one-third of the capital and subtract from it the share of a son. The remainder is one-third of the capital less the share. Then subtract from it one-fourth

of what remains of the one-third, namely, one-fourth of one-third less one-fourth of the share. The remainder is one-fourth of the capital less three-fourths of the share. Add hereto two-thirds of the capital : then you have eleven-twelfths of the capital less three-fourth of a share, equal to four shares. Reduce this by removing the three-fourths of the share from the capital, and adding it to the four shares. Then you have eleven-twelfths of the capital equal to four shares and three-fourths. Complete your capital by adding to the four shares and three-fourths one-fourth of the same. Then you have five shares and two-eleventh, equal to the capital.
(from which now each share and each legacy can be calculated.)

The solution appears to be very cumbersome; this is so because Al-Khowarizmi has used no notation. If he had, the solution would have run as under.

Let the property the dying man leaves behind be P; and let x be the ultimate share of each son. According to the terms of the legacy,

$$\text{the share of the first stranger} \quad = x$$

$$\text{and the share of the second stranger} \quad = \frac{1}{4}(\frac{P}{3} - x)$$

so that the four shares of the four sons together, namely, $4x$, the share x of the first legatee and the share $\frac{1}{4}(\frac{P}{3} - x)$ of the second legatee must together make up the capital.
That is,

$$4x + x + \frac{1}{4}(\frac{P}{3} - x) = P$$

which gives

$$\frac{19}{4}x = \frac{11}{12}P$$

and the share of each son is $\frac{11}{57}P$, from which we get

$$\text{share of the first legatee} = \tfrac{11}{57}P$$

$$\text{and share of the second legatee} = \tfrac{2}{57}P$$

9. Omar Khayyam.

The next algebraists from Baghdad to influence the growth of Algebra was the famous poet Omar Khayyam. His success as a poet who composed *Rubaiyat* has been so great that it has completely shrouded the public memory of his contribution to Algebra though this latter also was equally significant.

His main contribution is in respect of the

solution of the cubic equation.

To start with, he enumerated the cubic equations in seven distinct forms : namely,

$$\begin{aligned} x^3 &= b \\ x^3 + ax &= b \\ x^3 + b &= ax \\ x^3 &= ax + b \\ x^3 + ax^2 &= b \\ x^3 + b &= ax^2 \\ x^3 &= ax^2 + b \end{aligned}$$

equations which involve three terms. He solved each of them by showing that these solutions are points of intersection of two conic sections in each case, a parabola and a parabola or a parabola and a circle or a parabola and a hyperbola.

It was a time much before the discovery of analytic geometry. Hence Omar Khayyam drew his conics by the classical geometrical methods. We shall show his solutions by drawing the conic sections by using their standard equations as in analytical geometry.

The following is his solution of the cubic

$$x^3 = b$$

Draw the two parabolas

$$y = x^2 \quad \text{and} \quad y^2 = bx$$

on the same axes.

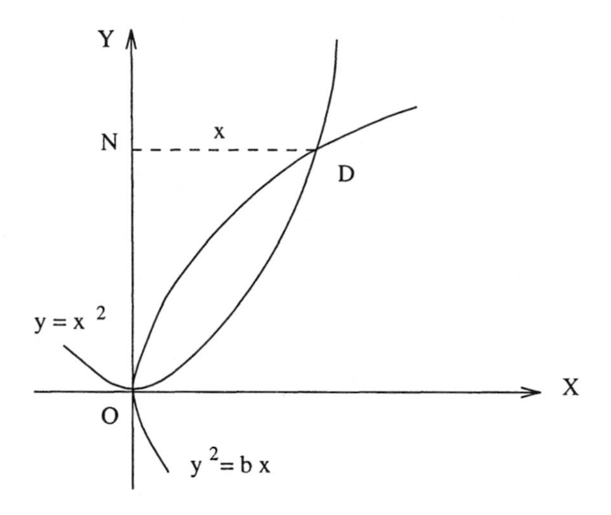

Let D with coordinates (x, y) be the point where the two parabolas intersect. Since D lies on the parabola $y = x^2$ and also on the parabola $x = y^2/b$, its x-coordinate x would satisfy

$$x = y^2/b = x^4/b$$

from which it follows that

$$x^3 = b$$

thus giving DN as the solution of the cubic $x^3 = b$.

For solving the cubic

$$x^3 + ax = b$$

which is his *second* form, Omar Khayyam uses

the parabola $y = x^2/\sqrt{a}$ and the circle $x^2 + y^2 = (b/a)x$.

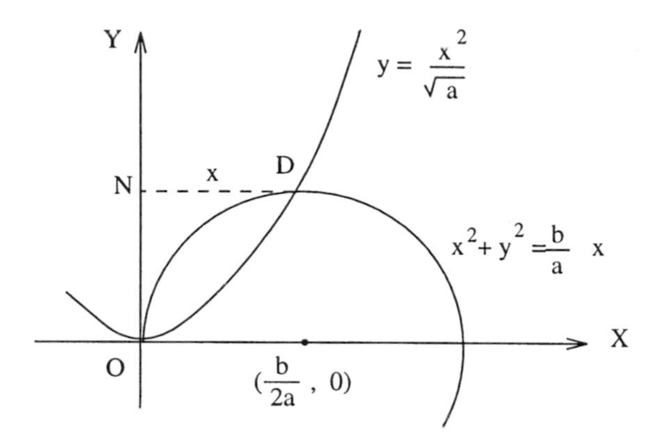

Let D(x, y) be the point where the two intersect. Combining the two facts that D lies on the circle as well as on the parabola, he gets x of D satisfying

$$x^2 + x^4/a = (b/a)x$$

from which it follows that $ax + x^3 = b$, or that $x^3 + ax = b$. This means that DN is the solution of the cubic

$$x^3 + ax = b.$$

For finding a solution of the cubic

$$x^3 + b = ax$$

which is his *third* form, Omar Khayyam uses

the parabola $y = x^2/\sqrt{a}$ and the hyperbola $x^2 - y^2 = (b/a)x$

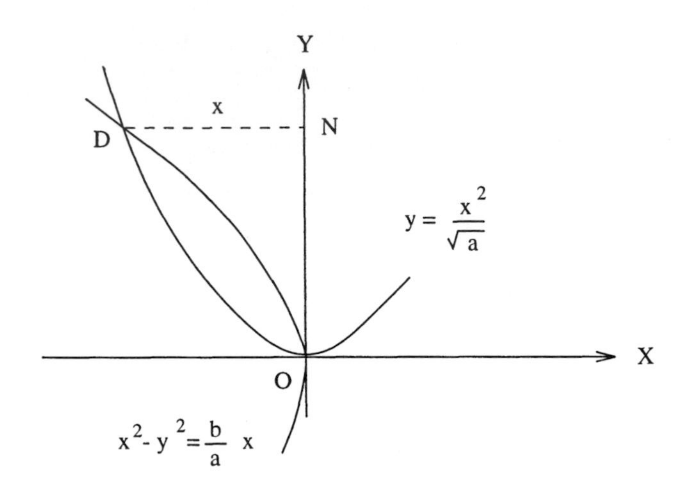

and uses the two facts that D(x, y) lies on $x^2 - y^2 = (b/a)x$ and on $y = x^2/\sqrt{a}$. Combining these two results, he gets

$$x^2 - x^4/a = (b/a)x$$

from which he gets

$$ax - x^3 = b$$

that is,

$$x^3 + b = ax.$$

Thus DN is the solution of the cubic $x^3 + b = ax$.

He next considers the cubic

$$x^3 = ax + b$$

which is his *fourth* form. To solve this, he uses

the parabola $y = x^2/\sqrt{a}$ and the hyperbola $y^2 - x^2 = (b/a)x$.

If D(x, y) is the point where the two intersect, one gets

$$x^4/a - x^2 = (b/a)x.$$

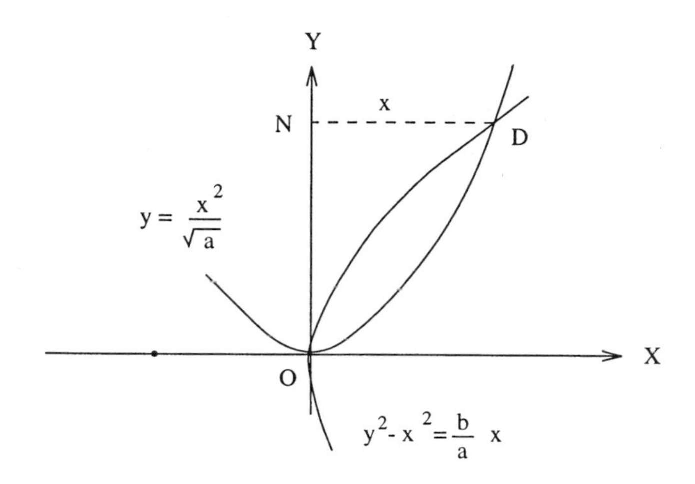

When simplified, this gives

$$x^3 - ax = b$$

that is,

$$x^3 = ax + b$$

so that DN is the solution of the cubic $x^3 = ax + b$.

Omar Khayyam's next cubic is

$$x^3 + ax^2 = b$$

which is his *fifth* form. To solve this, he uses

the parabola $y^2 = b^{1/3}(x + a)$ and the hyperbola $xy = b^{2/3}$.

If D(x, y) is the point where the two intersect, (x, y) satisfies both the equations. Combining these two results

$$x^2y^2 = b \cdot b^{1/3}$$

$$\text{and} \qquad y^2 = b^{1/3}(x + a)$$

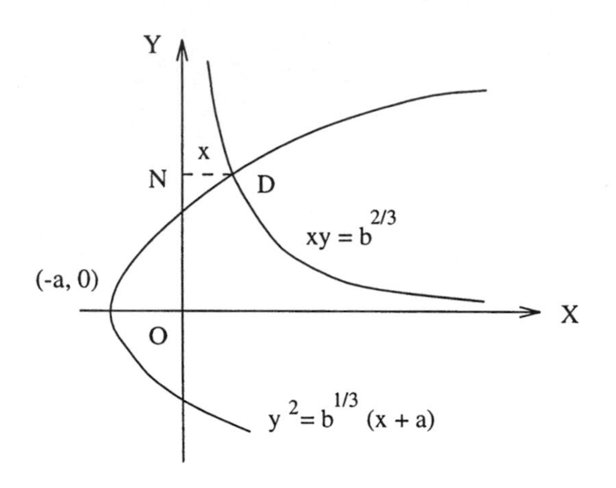

he gets

$$x^2 \cdot b^{1/3}(x+a) = b \cdot b^{1/3}$$

so that

$$x^3 + x^2 a \quad = b$$

$$\text{or} \qquad x^3 + ax^2 \quad = b.$$

Thus DN $= x$ is a solution of the cubic

$$x^3 + ax^2 = b.$$

Omar Khayyam next considers the cubic

$$x^3 + b = ax^2$$

which is his *sixth* form. To solve this he uses

the parabola $y^2 = -b^{1/3}(x-a)$ and the hyperbola $xy = -b^{2/3}$.

If D(x, y) is the point in which the two curves intersect, x and y satisfy both the equations

$$xy = -b^{2/3} \quad \text{and} \quad y^2 = -b^{1/3}(x-a).$$

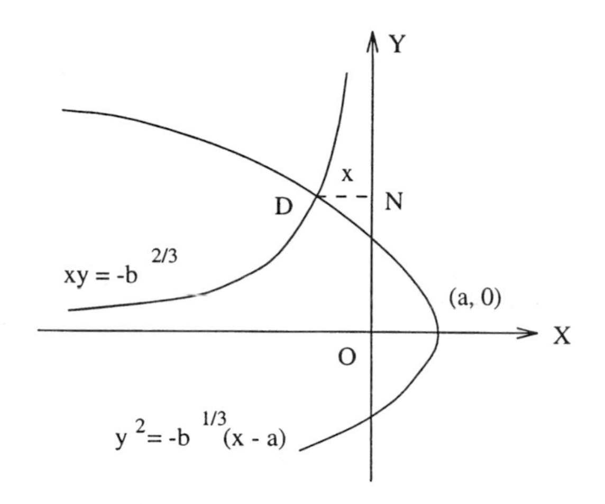

From these two he eliminates y and gets firstly

$$x^2y^2 = b^{4/3}$$

in which if the value of y^2 is put from the equation of the parabola, he would get

$$\begin{aligned} x^2[-b^{1/3}(x-a)] &= b \cdot b^{1/3} \\ \text{that is,} \quad x^2(x-a) &= -b \\ \text{that is,} \quad x^3 + b &= ax^2, \end{aligned}$$

the given cubic. Thus DN is the root of the cubic

$$x^3 + b = ax^2.$$

Finally, Omar Khayyam comes to the form

$$x^3 = ax^2 + b$$

which is the last and the *seventh* form of the cubics he enumerated at the start.

To solve this cubic, he employs

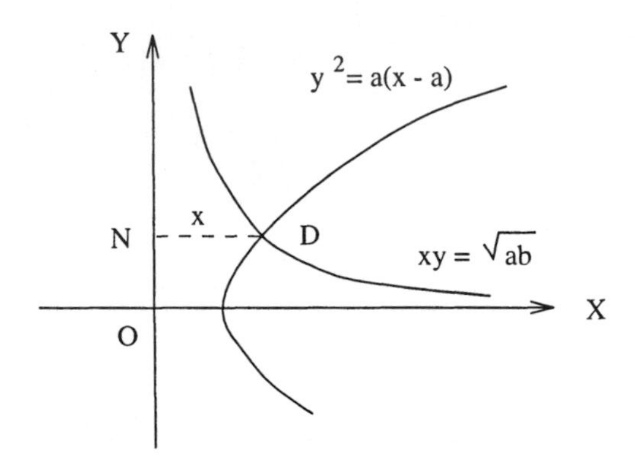

the parabola $y^2 = a(x-a)$ and the hyperbola $xy = \sqrt{ab}$.

At D(x, y) where the curves meet, x and y satisfy both the equations

$$y^2 \quad = a(x-a)$$

$$\text{and} \qquad xy \quad = \sqrt{ab}$$

so that from the two together he derives the result

$$\begin{aligned} ab = x^2y^2 &= x^2a(x-a) \\ \text{that is} \quad ab &= ax^3 - a^2x^2 \\ \text{which reduces to} \quad x^3 &= ax^2 + b \end{aligned}$$

showing that DN $= x$ is a solution of the cubic

$$x^3 \quad = ax^2 + b.$$

10. Liber Abbaci.

Omar Khayyam also did some more work of the same kind. But it did not have much of an impact on either the beginning of Algebra in Europe or on its growth. The work which brought Algebra to Europe was the treatise

Liber Abbaci

of Leonardo of Pisa (more commonly remembered as Fibonacci) who drew his inspiration and some of the ideas directly from the work of Al-Khowarizmi. This book of Fibonacci, the title of which means Book of Arithmetic, achieved three important results. Firstly, this was the book which was mainly instrumental in introducing Indian numerals and Indian arithmetical algorithms in Europe and thereby quickening the pace and ease of the computational side of mathematics. It took a few hundred years, though, for Europe to absorb and adopt this new arithmetic and its numerals. The major part of this final success, however, was undoubtedly that of this book of Fibonacci.

Fibonacci has devoted the second part of the book to posing and solving a number of interesting problems, some leading to linear equations and many to indeterminate linear equations. This enhances the importance of the book as the harbinger of a fresh interest in mathematics in Europe, from where it had almost disappeared for about twelve centuries.

Even where his problems are of a commercial nature, they differ from those of Al-Khowarizmi; which is natural : because problems arising from the social and mercantile practices depend essentially on the cultural and historical traditions of the region and the time. They cannot be the same in Italy in the early thirteenth century as they were in the eighth and ninth centuries in the region around Baghdad.

But there is a greater reason for the difference. Al-Khowarizmi had written more or less for practitioners in trade and law. Fibonacci, on the other hand, though the son of a merchant, had basically the make-up of an intellectual. Solutions of Al-Khowarizmi's six forms of quadratic equations did appeal to him; but not his problems of legacy and distribution of profits.

His interests made him go to problems of deeper mathematical excitement. Some problems which he chose to pose and the ingenious ways he employed to solve some of them deftly show why it is said that

With Leonardo a new epoch began in Western mathematics.

The following are some illustrations.

> A man buys 30 birds – partridges, doves and sparrows. A partridge cost him three silver coins, a dove two and a sparrow, a half. The total price he paid for the 30 birds was 30 silver coins. Find how many partridges, how many doves and how many sparrows he bought.

The problem, as can be seen, was to solve, the two equations

$$x + y + z \quad = 30$$

and

$$3x + 2y + \frac{1}{2}z \quad = 30$$

in the three unknowns x, y and z. It is a set of indeterminate equations. To propose and handle it at a time when there was no Algebra, no symbols, no notation and no established method was surely a remarkable step ahead.

Another similar problem is the following

> One of two persons says to the other: "If you can give me one third of your cash, I can buy that horse." The other then says to the first "If you can give me one quarter of your cash, I can buy that horse." Find the cash each had and the probable price of the horse.

This also, as one can see, is the problem of solving the two equations

$$x + \frac{1}{3}y \quad = s$$

and

$$y + \frac{1}{4}x \quad = s$$

and find the three unknowns x, y and s.

There are many other similar problems in the book. The most outstanding and extremely original problem which Fibonacci had included in the book is the following one, known as

Fibonacci's Rabbit Problem

about the number of pairs of rabbits which can be bred from one pair in a year's time: namely

> A man has a pair of rabbits at a place entirely surrounded by a wall. The nature of the rabbits is such that they breed every month one other pair, which begins to breed in the second month after their birth. Find how many pairs would have been bred in a year's time.

Since it is a question of calculating the number in just twelve months, one can do it even by trial and error and find that the number of rabbit-pairs month wise is

$$1, 2, 3, 5, 8, 13, 21, 34, 55, 89, 144, 233$$

This series has the characteristic that every term after the second is the sum of the previous two terms. Because of this characteristic the series has come to stay in Mathematical literature ever since and is known as

Fibonacci Series.

It has become a subject of further research in Algebra which has led to the discovery of a number of interesting results.

As in constructing or in choosing the problems, so in the solutions of some of them, Fibonacci showed very pleasing ingenuity. We give two illustrations. One is his solution of the system

$$\begin{aligned} & x + \frac{1}{3}(y+z) &= s \\ & y + \frac{1}{4}(z+x) &= s \\ \text{and} \qquad & z + \frac{1}{5}(x+y) &= s \end{aligned}$$

of 3 equations in 4 unknowns.

To solve the system, he introduces an auxiliary unknown

$$t = x + y + z$$

Subtracting each of the three give equations from the last one, he obtains

$$\begin{aligned} \frac{2}{3}(y+z) &= t-s \\ \frac{3}{4}(z+x) &= t-s \\ \text{and} \qquad \frac{4}{5}(x+y) &= t-s \end{aligned}$$

Writing D for $t-s$, he puts them in the forms

$$\begin{aligned} y+z &= \tfrac{3}{2}D \\ z+x &= \tfrac{4}{3}D \\ \text{and} \qquad x+y &= \tfrac{5}{4}D \end{aligned}$$

Looking at the denominators of D on the right, he puts $D = 24$ so that he could get integer solutions of the system. This device leads him to the set

$$\begin{aligned} y+z &= 36 \\ z+x &= 32 \\ \text{and} \qquad x+y &= 30 \end{aligned}$$

From which he gets the final solution

$$x = 13, \quad y = 17 \ \text{ and } \ z = 19$$

and using these values of x, y and z he obtains $s = 25$.

It is surprising that about 800 years ago, when Algebra was just in infancy, Fibonacci should think of introducing an auxiliary unknown to solve a given system of linear equations.

Fibonacci exhibited still greater ingenuity in solving the problem

> Find a rational number x such that $x^2 + 5$ as well as $x^2 - 5$ is a square.

This problem was set to him as a challenge to his mathematical abilities by Giovanni da Palermo, a courtesan of Fredrik II. The problem amounts to finding x from a pair of equations

$$x^2 + 5 \quad = y^2,$$

$$x^2 - 5 \quad = z^2.$$

A little thinking convinced him that no integers would satisfy the pair. He therefore gave it the more general form of finding x, y, z from the pair

$$x^2 + c = y^2$$

$$x^2 - c = y^2$$

so that no special property of number 5 may obstruct the way of noticing the real intrinsic difficulties of the situation.

He now adds these two latter equations and gets

$$2x^2 = y^2 + z^2$$

a form, a little closer to those which we can simplify by the use of standard results. To transform it to a still simpler form, he puts

$$y = u + v, \quad z = u - v$$

and gets the last equation simplified to the form

$$x^2 = u^2 + v^2.$$

This is the equation for Pythagorean triplets. Fibonacci must have been familiar with its solutions, one part of which is

If a and b are *any* relatively prime odd integers, numbers

$$x = \frac{a^2 + b^2}{2}, \; u = ab, \; v = \frac{a^2 - b^2}{2}$$

is a solution of

$$x^2 = u^2 + v^2.$$

Now from the two equations, $x^2 + c = y^2$ and $x^2 - c = z^2$ he gets

$$2c = y^2 - z^2 = (u+v)^2 - (u-v)^2 = 4uv$$

in which he puts the values of u and v as above and gets

$$2c = 4ab\frac{a^2 - b^2}{2}$$

so that

$$c = ab(a^2 - b^2)$$

where a and b are any two relatively prime odd integers. In order to get a convenient value for c, he chooses $a = 9$, $b = 1$ and gets

$$c = 9 \cdot 1 \cdot (9^2 - 1^2) = 9 \cdot 80 = 720 = 5 \times 12^2.$$

For the values $a = 9$, $b = 1$, he gets

$$x = 41,\ x^2 + c = 41^2 + 720 = 49^2,\ x^2 - c = 41^2 - 720 = 31^2.$$

Dividing by 12^2, he gets

$$\left(\frac{41}{12}\right)^2 + 5 = \left(\frac{49}{12}\right)^2, \quad \left(\frac{41}{12}\right)^2 - 5 = \left(\frac{31}{12}\right)^2$$

and thus gets $(41/12)$ as the solution of Giovanni's problem.

11. The next stage.

Fibonacci and his work found a number of followers among his contemporaries and immediate successors. Many of them extended his work, but none broke very much of new ground. But, together, they did an important service to mathematics. They kept alive the cultivation of Algebra.

They showed for Algebra the enthusiasm and warmth which, once a thousand years earlier, Greece and Greek mathematicians had shown for Geometry. They brought to Algebra a status similar to, but not as high though, as that which Greece and Alexandria brought to Geometry. The intellectual climate of Greece, then, was such that if anybody wanted to do mathematics or follow any other intellectual pursuit, he was compelled to study and discover new results in Geometry. This was exactly the status which Algebra acquired during the two to three hundred years following Fibonacci's time.

Till Fibonacci's time, the quadratic equation was thoroughly studied in all its six forms set by Al-Khowarizmi and included by Fibonacci in his *Liber Abbaci.*

This book and its contents, in a way, set a direction for work in Algebra. Since the problem of the quadratic equation was completely solved and a large number of interesting applications found for the quadratic equation, mathematicians who now came in the area considered the algebraic solution of cubics and other equations of higher degree to be the next natural attempt to make. They set themselves, therefore, to the task of attempting the solutions of such equations of degree higher than two. They also thought it natural to have these solutions in terms of radicals since the solution of the quadratic equation was of such nature.

12. The Cubic.

This work was undertaken and completed successfully in the sixteenth century by a group of four Italians: Scipione del Ferro, Niccolo Fontana (known generally as Tartaglia even in the mathematical literature because he was a stammerer), Gerolame Cardano and Ludovico Ferrari. We have grouped them together because they were all together involved in the solutions of the cubics and the quartics. But actually, apart from this one respect that they were all involved in the same mathematical work, there was very little common between them. They came from different backgrounds and belonged to different age-groups. Ferro was a professor at the university of Bologna while Ferrari was a house-hold servant. When Ferro died in 1526 at the age of sixty one, Ferrari was just four years old. Tartaglia was a teacher in Venice and was an extremely simple gentleman. Cardano, on the other hand, had a weakness for gambling, who had however made a virtue of this weakness. He wrote a book on the mathematics of gambling and the stakes involved therein. This book happens to be the first book on the theory of Probability. To the biography of this colorful personality which Oystein Ore wrote, he gave the title *The Gambler Scholar*, an apt title indeed.

13. The Cubic $x^3 + ax = b$.

It seems that the first to solve this cubic equation was Ferro. But as was usual in those times, he did not part with his solution nor give it any publicity. In the last years of his life, he shared his secret with a younger mathematician, Antoni Fiore, a favorite student of his. Fiore also, for a long time, kept it away from the general public.

Cardano was the first to publish its solution.

In order that the method be properly understood, he illustrated the same by applying it to solve the equation

$$x^3 + 6x = 20.$$

To solve this, he set

$$x = u - v$$

incidentally also suggesting that the same step has to be taken for all similar equations. With this substitution, he obtained

$$\begin{aligned} x^3 + 6x &= (u-v)^3 + 6(u-v) \\ &= (u^3 - v^3) - 3uv(u-v) + 6(u-v) \end{aligned}$$

and thus he had his original equation turned into

$$(u^3 - v^3) - (u-v)(3uv - 6) = 20$$

He now chose the two auxiliary unknowns u, v to satisfy the two conditions

$$\begin{aligned} u^3 - v^3 &= 20 \\ 3uv &= 6 \end{aligned}$$

From $3uv = 6$, he obtained $uv = 2$ and $u^3v^3 = 8$ and from the two conditions

$$u^3 - v^3 = 20$$

$$u^3v^3 = 8$$

he obtained

$$\begin{aligned}(u^3 + v^3)^2 &= (u^3 - v^3)^2 + 4u^3v^3 \\ &= 400 + 32 \\ &= 432 = 4 \times 108 \\ \text{so that} \quad u^3 + v^3 &= 2 \times \sqrt{108}\end{aligned}$$

Combining this with

$$u^3 - v^3 = 20$$

he obtained

$$u^3 = 10 + \sqrt{108}$$

$$v^3 = -10 + \sqrt{108}$$

so that, since $x = u - v$, he obtained

$$\sqrt[3]{10 + \sqrt{108}} - \sqrt[3]{-10 + \sqrt{108}}$$

as a solution of the original cubic, $x^3 + 6x = 20$. Being also an expression in radicals this completed the task of solving the cubic of the form $x^3 + ax = b$ in terms of radicals.

Cardano also formulates the general rule in this respect. In his book *Ars Magna* in which he published this solution, the rule reads as under.

> Cube one third the number of sides (i.e., one third of the coefficient of x). Add to it the square of one half the

constant of the equation, and take the square-root of the whole. You will put this twice, and to one of the two you add one-half the number you squared and from the other you subtract one half the same. Then subtracting the cube-root of the second from the cube-root of the first, you get the required solution.

In our present language and terminology, this would read as follows.

The solution of the cubic equation

$$x^3 + ax = b$$

is

$$\sqrt[3]{\frac{b}{2} + \sqrt{\frac{b^2}{4} + \frac{a^3}{27}}} - \sqrt[3]{\frac{-b}{2} + \sqrt{\frac{b^2}{4} + \frac{a^3}{27}}}.$$

This solution of a cubic given by Cardano in his book *Ars Magna* printed in 1545 is, indeed, a clever one. But it is not originally his. He got it first from Tartaglia and also saw it later in Ferro's work.

The story of how Cardano got it from Tartaglia is interesting.

In those days it was customary amongst mathematicians in Italy to challenge each other in problem-solving contests. Of the two mathematicians participating in such a contest, each was supposed to set a certain number of problems for the other one to solve. It appears that Tartaglia had a rather extraordinary skill in setting as well as in solving knotty problems. He had won a number of such contests. It once happened that he was challenged to such a contest by a young mathematician Fiore, a student of the great departed Professor Ferro. Tartaglia accepted the challenge. Each was supposed to set 30 problems for the other to solve. Fiore was a favorite student of Professor Ferro and had come to know the method to solve cubics of the type $x^3 + ax = b$ from his teacher who till then was the only one to know the method. That was the greatest weapon in his armory and he knew that if he set Tartaglia problems leading to such cubics, Tartaglia would lose and he would win. Accordingly he set all 30 problems of that single type. When faced with them Tartaglia was a little surprised and he could guess that there must have been a method to solve such cubics and that Fiore must have been knowing it. Since Tartaglia felt sure that a method for it existed, he worked his hardest on those two days of the contest and *succeeded* in discovering a method to solve these cubics. He solved all the problems set by Fiore and won the award. Tartaglia was very very happy with this discovery and the contest which gave him the opportunity to make it.

This was in 1535. Though not the discovery, the news that Tartaglia had made the discovery reached many. Cardano was one of them. And he was very keen to learn it from Tartaglia. But Tartaglia would not oblige. Finally Cardano lured Tartaglia to come over to Milan. Cardano promised to introduce him to the marchese Alfonso d'Avalos, the military commander of Milan. Tartaglia had made some military inventions and was keen to bring them to the notice of the marchese. He therefore accepted Cardano's invitation and came over to Milan as a guest of Cardano. Cardano must have introduced him to the marchese and Tartaglia must have felt extremely obliged to Cardano for the same. Under the weight of this obligation Tartaglia finally agreed to the request of Cardano. He showed his method of solving a cubic of type $x^3 + ax = b$ to Cardano under a promise which Cardano gave that he would not publish it. Cardano even took the following solemn oath:

> I swear to you by the sacred Gospel, and on my faith as a gentleman, not only never to publish your discoveries, if you tell them to me, but I also promise and pledge my faith as a true Christian to put them down in cipher so that after my death no one shall be able to understand them.

This was on March 25, 1539. Cardano faithfully kept his promise. In the meanwhile, another intriguing development took place. In 1536, a boy of 14, by name Ludovico Ferrari had approached Cardano for work. Cardano took him as a servant in the household. It had soon become clear to Cardano that Ferrari was extraordinarily precocious. As William Dunham puts it in his book *Journey*

Through Genius,

> Their relationship quickly turned from master-servant to teacher-pupil and eventually, before Ferrari was 20 years old, to colleague-colleague. Cardano shared Tartaglia's secret with his brilliant young protege and together the two of them made astounding progress.
>
> Meanwhile, Ferrari succeeded in finding a technique for solving the quartic equation. This was a major discovery in algebra; but it depended on reducing the quartic to a related cubic and again Cardano's oath forbade its publication. The two men possessing the greatest algebraic discoveries of their time were stymied.

The tension of the situation was too much. They wanted to wriggle out of it. In 1543, they traveled to Bologna where they had hoped to be able to inspect the papers of Ferro and had expected to find amongst them some evidence that Ferro had already discovered the method. And there it really was; in Ferro's own hand: a solution of the cubic $x^3 + ax = b$. Cardano felt so relieved. He argued to himself that if he published it now, he was publishing Ferro's solution, not that of Tartaglia. This slender argument appeared to him a sufficient moral justification for publishing it and yet claiming that he had not broken his solemn pledge to Tartaglia.

And in 1545, in his book *Ars Magna* the solution was published in Chapter XI of the book. It is worth noting that Cardano begins this chapter with the following words of explanation:

> Scipio Ferro of Bologna well-nigh thirty years ago discovered this rule and handed it on to Antonio Maria Fiore of Venice whose contest with Niccolo Tartaglia of Brescia gave Niccolo occasion to discover it. He *gave it to me in response to my entreaties*, though withholding the demonstration. Armed with this assistance, I sought out its demonstration in various forms. This was very difficult.

Cardano had thus given credit where credit was due.

14. The Cubics $x^3 + ax = b$ and $x^3 + b = ax$.

Tartaglia had visited Cardano in 1539 and had passed on to him the secret of the technique of solving $x^3 + ax = b$. It consisted of introducing two auxiliary unknowns u and v and imposing on them two conditions so that $u - v$ could be a root of the cubic. With the help of these conditions u, v and $u - v$ could be determined.

But this would not work for cubics $x^3 = ax + b$ and $x^3 + b = ax$. But Cardano was clever enough to think of alternatives. He found that assuming

$$u + v$$

as a solution and using the same technique as earlier, u and v could be determined and so a solution could be found.

But now an altogether different difficulty arose. He took the equation

$$x^3 = 15x + 4$$

for an illustration. Putting

$$x = u + v$$

in this led him to

$$(u + v)^3 = 15(u + v) + 4$$

that is, to

$$u^3 + v^3 + 3uv(u + v) = 15(u + v) + 4$$

If u and v are subjected to the conditions

$$u^3 + v^3 \quad = 4$$

$$3uv \quad = 15$$

and u, v determined as in the case of $x^3 + ax = b$, one obtains

$$(u^3 - v^3)^2 \quad = (u^3 + v^3)^2 - 4u^3v^3$$

$$= 16 - 500 = -484$$

a square on the left and a negative number on the right; a situation impossible to accept.

Cardano, however, pushed on and *formally* wrote

$$u^3 + v^3 = 4$$

$$u^3 - v^3 = \sqrt{-484}$$

and obtained

$$u^3 = 2 + \frac{1}{2}\sqrt{-484}, \quad v^3 = 2 - \frac{1}{2}\sqrt{-484}$$

and

$$\sqrt[3]{2 + \frac{1}{2}\sqrt{-484}} + \sqrt[3]{2 - \frac{1}{2}\sqrt{-484}}$$

as the solution of the cubic $x^3 = 15x + 4$.

Now this was very embarrassing. Still Cardano pushed on. He took the square root $\sqrt{-484}$ as if it was a real number and by the application of the usual rules of real algebra, brought it to its simplest form. Since 484 is the square of 22, this simplest form turned out to be

$$22 \cdot \sqrt{-1}$$

and the expression

$$\sqrt[3]{2 + 11\sqrt{-1}} + \sqrt[3]{2 - 11\sqrt{-1}}$$

turned out to be the solution of the cubic.

Cardano did not know how to find the cube-roots involved or even whether these cube-roots exist and if they exist, whether they have any meaning.

By repeated trials, Cardano found that

$$2 + 11\sqrt{-1} \text{ was the cube of } 2 + \sqrt{-1}$$

and

$$2 - 11\sqrt{-1} \text{ was the cube of } 2 - \sqrt{-1}$$

and consequently, the solution of the cubic to be

$$(2+\sqrt{-1})+(2-\sqrt{-1})$$

that is, the *real number* 4.

On verification, he found that 4 was, indeed, a solution of the equation $x^3 = 15x + 4$.

This was extremely intriguing. You start with a perfectly good looking equation with simple positive integers as coefficients. In the intermediate stages of the work you come across horrid-looking expressions like $\sqrt{-484}$ and when you push on applying to it, with absolutely no justification, the algebraic rules of real numbers and simplify, you finally end again with a perfectly acceptable positive integer like 4 as your answer.

You and I know today that the two terms $2 + 11\sqrt{-1}$ and $2 - 11\sqrt{-1}$ are conjugate complex numbers. But Cardano did not. Nor any one else in his time. Cardano was possibly the first who significantly came face to face with complex numbers. He was not very happy with them. But still he did not avoid them. And, what is most fortunate is that he was not shy of them. He continued to work with them. And this work considerably helped create a climate in the mathematical world in which they gradually acquired increasing acceptability. And finally they came to stay in mathematics for good and play in it a useful role.

15. The Quartic.

In 1540, Zuanne de Tonini da Coi, an Italian mathematician posed the following problem to Cardano:

> Divide 10 into three parts such that they will be in continued proportion and the product of the first two will be equal to 6.

If the three parts are denoted by x, y, z, Coi's problem reduces to finding x, y, z from the three equations

$$x+y+z=10, \quad xz=y^2, \quad xy=6$$

Eliminating x and z from the first equation with the help of the other two, we get

$$\frac{6}{y} + y + \frac{y^2}{6/y} = 10$$

or

$$y^4 + 36 + 6y^2 = 60y$$

This was a quartic. Cardano could not solve it. But his student Ferrari did and in the process gave a method to solve a general quartic.

Ferrari wrote the equation in the form

$$x^4 + 6x^2 + 36 = 60x$$

added $6x^2$ to both sides to obtain

$$x^4 + 12x^2 + 36 = 6x^2 + 60x$$

That is

$$(x^2 + 6)^2 = 6x^2 + 60x$$

If the expression on the right at this stage had turned out to be a square P^2 of an expression P, the equation would have been

$$(x^2 + 6)^2 = P^2$$

and could have led to the two quadratic equations

$$x^2 + 6 = P, \quad x^2 + 6 = -P$$

which could have been both solved to give us the four roots of the original quartic.

But as it was, the expression on the right was not a square. Ferrari, therefore, added a number b within the bracket on the left and made it

$$(x^2 + 6 + b)^2$$

which means he had added $2bx^2 + 12b + b^2$ to the left side. He therefore added the same on the right and obtained

$$\begin{aligned}(x^2+6+b)^2 &= 6x^2+60x+2bx^2+12b+b^2\\ &= (6+2b)x^2+60x+(12b+b^2) \qquad (1)\end{aligned}$$

Now, he argued that this right side would be a square if

$$(6+2b)(12b+b^2) = 900$$

or if

$$b^3+15b^2+36b = 450$$

Now this is a cubic and they could solve it. If

$$b = p$$

is the solution, it would mean that for this value p of b the right side expression of equation (1) would be a square : a square of

$$\sqrt{6+2p}\,x+\sqrt{12p+p^2}$$

so that equation 1 would be of the form

$$(x^2+6+p)^2 = (\sqrt{6+2p}\,x+\sqrt{12p+p^2})^2$$

from which one could now obtain the two quadratic equations

$$x^2+6+p = \sqrt{6+2p}\,x+\sqrt{12p+p^2}$$

$$x^2+6+p = -\sqrt{6+2p}\,x-\sqrt{12p+p^2}$$

Solving these two equations, one could obtain all the four roots of the original quartic.

Cardano has included this discovery of Ferrari in his *Ars Magna* and has very specifically mentioned in the book that this is a discovery of Ferrari and that Ferrari gave it to Cardano on the latter's request for the same.

16. A look-back.

The Italian group of Ferro, Tartaglia, Cardano and the young Ferrari had done an excellent job of the task which they had set for themselves of solving the cubic and the quartic in radicals. When we now look back on this work of theirs we are struck that there is an interesting uniformity of pattern in it. The cubic was solved with the help of a quadratic and to solve a quartic assistance was taken of a cubic which they solved as a preliminary step. Thus, in solving the cubic $x^3 + 6x = 20$, they introduced two auxiliary unknowns u, v which satisfied the conditions

$$u^3 - v^3 = 20$$

$$uv = 2$$

which together led to

$$u^6 - 20u^3 - 8 = 0$$

a quadratic in u^3.

Very similarly, for solving the quartic $x^4 + 6x^2 + 36 = 60x$, they first found out an auxiliary unknown b from a cubic

$$b^3 + 15b + 36b = 450.$$

The lesson implicit in the uniformity of pattern of these derivations was quite clear. Ferrari could not have missed it. He was only 20 when he had discovered the solution of the quartic. He was feeling sure that with a whole life before him, he would be able to solve equations of degrees higher than 4.

But he could not.

Nor could so many who followed. Mathematicians who have added to Algebra so much other important material failed in respect of this problem.

They must have finally suspected that the problem concerned is not solvable. But they could not prove this either. And this disturbing state of animated suspense continued until it was resolved by two young geniuses : one from Norway and the other from France. The one from Norway was Abel – Neil Henrik Abel; and the other was Galois – Evariste Galois from France. It is a strange coincidence that both of them died when young – Abel at 27 and Galois at 20 : one of tuberculosis and the other in a duel.

The crisis they solved was truly intriguing. It is strange that a procedure should succeed for equations of degrees 2, 3 and 4 and should suddenly cease to succeed for equations of degree 5. Was it so because numbers 2, 3 and 4 possess some peculiar property which number 5 does not possess? Or is it for some other inexplicable reason?

Abel was the first to resolve this question : Galois the next. But each independently : each his own way. They both in a way established the same general conditions for equations to be solvable in radicals.

In Galois's language, the criterion is:

> An equation is solvable by radicals if and only if its Galois group G is solvable; that is, if G possesses the composition series
>
> $$\mathrm{G} \supset \mathrm{H}_1 \supset \mathrm{H}_2 \supset \cdots \supset \mathrm{H}_m = \mathrm{E}$$
>
> in which all indices are prime numbers.

17. The new era.

With this, one era was over for Algebra. In this era what was mainly investigated was *equations* – equations in one variable but of different degrees, systems of linear equations, indeterminate equations of the first and second degrees and cyclotomic equations.

With Galois's theorem the very face of Algebra changed. Equations were no longer its main center of investigation. The central theme of the Algebra was now

algebraic structures,

Groups, Fields, Rings and various other similar abstract structures.

Galois's theorem had, in a way, unified all previous major work on equations. Realization of this strength of abstract generalization had come only to some. And they were vigorously constructing more and more abstract structures to suit and unify the variety of situations arising in mathematics and mathematical sciences. Non-commutative algebras, non-associative algebras, Lie algebras, Jordan algebras were all created as and when need for them had arisen. Algebra, in this new era, consisted of these investigations.

But even then and concurrently with all this, the earlier ways of looking at mathematics, the earlier ways of mathematical thinking and the earlier ways of working out mathematical problems had continued exactly as of old.

The following episode brings out this difference in attitude between the old guard and the new enthusiasts for abstractions. Paul Gordan was a professor of mathematics at the Erlangen University in the late nineteenth century when Hilbert was professor at Gottingen. When Gorden retired from his post, Max Noether, a colleague wrote about him what is known as a "characterization," the last sentence of which was

Er war ein Algorithmiker.

Gordan had been the first to solve one of the major problems of Invariant Theory. Like an *Algorithmiker* he solved a part of it by working out all the invariants of ternary quadratic forms, of ternary cubic forms and of others. But he did not succeed in doing it for binary forms.

This was in 1868.

In 1888, Hilbert proved his theorem on the topic : he did not calculate the invariants, but proved by using abstract theoretical methods that the invariants concerned *exist.* When Gordan read the paper of Hilbert he remarked:

"Das ist nicht Mathematik : das ist Theologie,"

showing thereby his mistrust in abstract proofs of *existence.* He demanded the actual construction and demonstration of these invariants for showing that they do exist. This shows the difference between the mathematical thinking of Gordan and Hilbert. Gordan was an Algorithmiker, Hilbert an Abstractionist.

A story which throws further light on this emerging new attitude is told about Emmy Noether. Emmy Noether had finished her

Doctorate at Erlangen under the guidance of Gordan. The subject of her thesis was: *Complete Systems of Invariants for Ternary Biquadratic Forms.* This was done naturally in the spirit of Gordan's work. In later years, when she was lecturing at Gottingen, her attitude to mathematics was so much changed and she had become so much of an abstractionist that if ever the question of her doctoral thesis came up, she would simply brush away the question with a brisk reply: "Oh! Don't ask about it : that was not mathematics."

A little after 1917, Klein and Hilbert at Gottingen got intensely interested in Einstein's Theory of Relativity. By that time, Emmy Noether's interests also were centered on a connection between Invariant Theory and Relativity Theory. She gave at that time a genuine and universal mathematical formulation for two of the most significant ideas of Relativity.

When a copy of this paper was received by Einstein, he was simply overwhelmed. In a letter dated May 24, 1918, Einstein wrote to Hilbert:

> "Yesterday, I received from Miss Noether a very interesting paper on Invariant Forms. I am impressed that one can comprehend these matters from so general a point of view. It would not have done the Old Guard at Gottingen any harm, had they picked up a thing or two from her. She certainly knows what she is doing."

She certainly did know what she was doing. Her view point of mathematics was entirely a new one. With her and because of her Algebra once again had its face completely changed and a *new era started*, an era which can be rightly called the *era of Emmy Noether.*

In the obituary which Van der Waerden wrote after the death of Emmy Noether, he described the characteristics of the algebraic thinking of Emmy Noether and indirectly of the algebra of the new era which started with her and because of her:

> One could formulate the maxim by which Emmy Noether let herself be guided, as follows: *all relations between numbers, functions and operations, become clear, generalizable and truly fruitful, only when they are separated*

> *from their particular objects and reduced to general concepts.* For her, this guiding principle was by no means a result of her experience with the importance of scientific methods, but an *a priori* fundamental principle of her thoughts. She could conceive and assimilate no theorem or proof before it had been abstracted and thus made clear in her mind. She could think only in concepts, not in formulas, and this is exactly where her strength lay. In this way, she was forced by her own nature to discover those concepts that were suitable to serve as bases of mathematical discoveries.

Van der Waerden enunciated this as the principle of Emmy Noether's work. Broadly speaking, this was also the principal character of the Algebra of this new era.

It was unfortunate that Emmy Noether could not continue at Gottingen for long. In 1933 she shifted to the United States. But even there she could not work long. She died in 1935 at the age of 53 in an operation for a tumor. This end came too early and too suddenly for the *chief architect* of the new era in Algebra.

When she died Einstein wrote a letter to New York Times giving his estimate of Emmy Noether's influence within the totality of human experience. On May 4, 1935 New York Times published it on its front page. Amongst other things, Einstein wrote:

> The efforts of most human beings are consumed in the struggle for their daily bread; but most of those who are, either through fortune or some special gift, relieved of this struggle are largely absorbed in further improving their worldly lot. Beneath the effort directed towards the accumulation of worldly goods lies all too frequently the illusion that this is the most substantial and desirable end to be achieved; but there is, fortunately, a minority composed of those who recognize early in their lives that the most beautiful and satisfying experiences open to human-kind are not derived from the outside, but are bound up with the development of the individual's own feeling, thinking and acting. The gen-

> uine artists, investigators and thinkers have always been persons of this kind. However inconspicuously the life of these individuals runs its course, none the less the fruits of their endeavors are the most valuable contributions which one generation can make to its successors.

This is a tribute one great paid to another great : Albert Einstein to Emmy Noether : both keen that science should grow : both keen that mathematics should grow : both lending their able hands to the growth of

MATHEMATICS
THE MOST EXALTED PURSUIT OF
HUMAN INTELLECT.

Exercises

Exercises 1 - 5 are from Al-Jabr wul Al Muqabala of Al-Khowarizmi.

1. I have divided ten into two parts, and having multiplied each part by itself, I have subtracted the smaller from the greater and the remainder was forty. Find the two parts.

2. I have divided ten into two parts, and having multiplied each part by itself, I have put them together and have added to them what was the differences of the two parts before their multiplication, and the amount of all this is fifty four. Find the two parts.

3. There is a triangular piece of land, two of its sides having ten yards each and the basis twelve. What must be the length of one side of a square inscribed in such a triangle?

4. A man dies and leaves four sons and bequeathes to some person as much as the share of one of his sons; and to another, one fourth of what remains after the deduction of the above share from one third. Find the share of each.

5. A person dies and leaves two sons and two daughters behind; and bequeathes to some person as much as the share of a

daughter less one fifth of what remains from one third after the deduction of that share; and to another person as much as the share of the other daughter less one third of what remains from one third after the deduction of all this; and to another person half one sixth of his entire capital. Find the share of each.

[Note: A son's share is twice a daughter's share.]

6. Solve completely the problem of 30 birds given in section 10 of the Chapter.

7. A person buys 100 birds of three types, partridges, doves and sparrows. A partridge costs him 4 silver coins, a dove, 3 silver coins and a sparrow, one-fifth of a coin. If the 100 birds cost him 100 silver coins, find how many birds he purchased of each type.

8. If $u_1, u_2, u_3, \ldots$ is a Fibonacci series, show that $u_9 = 13u_1 + 21u_2$.

9. Find a rational number x such that $x^2 + 21$ and $x^2 - 21$ are both squares.

10. Find a rational number x such that $x^2 + 30$ and $x^2 - 30$ are both squares.

11. Apply the method which Omar Khayyam employed for his seventh form and find a root of $x^3 = 2x^2 + 9$.

12. Use the Tartaglia-Cardano method and find a root of $x^3 + 3x = 36$. Reduce it to its simplest form.

    [Hint: $18 + 5\sqrt{13} = (\sqrt{13}/2 + 3/2)^3$.]

13. Use Cardano's method and find a root of $x^3 + 8 = 18x$. Reduce the root to its simplest form.

14. Apply Ferrari's method to the equation $x^4 + 8x = 4x^2 + 224$ and show that the associated cubic is

$$b^3 + 8b^2 + 244b = 8.$$

15. Show that G = {1, 3, 9, 27} is a group under the operation of multiplication modulo 80.

16. Show that H= {0, 1, 2, 3} is a group under the operation of addition modulo 4.

17. Show that the multiplicative group G of Ex. 15 is isomorphic to the additive group H of Ex. 16.

Further Reading

1. Frederic Rosan, Ed. and Tr. *The Algebra of Mohammed Ben Musa*, The Oriental Translation Fund, London, 1831.

2. B. L. van der Waerden, *A History of Algebra From Al-Khwarizmi to Emmy Noether*, Springer-Verlag, Berlin, 1985.

3. William Dunham, *Journey through Genius – The great Theorems of Mathematics*, John Wiley & Sons, Inc., New York, 1990.

4. James Brewers and Martha Smith, Ed., *Emmy Noether – A tribute to her life and work*, Ed. Marcel Dekkar Inc., New York, 1981.

ANSWERS : HINTS : SOLUTIONS

Chapter I.

2. (c) Egyptian formula gives $\pi = 3.1605$;
   calculator gives $\pi = 3.1416$.

6. (a) g.c.d. = 11 (b) g.c.d. = 51
   (c) **Hint:** Show that their g.c.d. = 1.

7. $2^3 \times 3^2 \times 5 \times 7^2 \times 11$

9. 9, 7, 4, 11

10. $x = 177/16$ is one solution.

11. $x = 921/64$ is one solution.

Chapter II.

1. 726437; 124281; 734088

2.

3. MDCCCCLXXXXIII; MI

4. (a) CLXXXXVII (b) CCCCXXXV (c) DCCCCLXXXXII; 992

5. (a) MCCCLXIIII (b) CCCCLXXXIIII (c) MDCCXXVIII

6. (a) $s_4 = 890, \quad s_6 = 1315$
   (b) $\frac{s_4}{3438} = 0.2589 \quad \sin 15° = 0.2588$

   $\frac{s_6}{3438} = 0.3825 \quad \sin 22°30' = 0.3826$

Chapter III.

1. **Hint:** Midpoint P of AB is (25, 4); therefore C is (16, 4).

$$\text{Area (triangle ADC)} = \frac{1}{2}\begin{vmatrix} 1 & 49 & 7 \\ 1 & 16 & 4 \\ 1 & 1 & 1 \end{vmatrix}.$$

3. 6^5.

4. $81/4$.

5. $16/3$.

6. $64/5$.

7. 12.

8. (a) $38/3$ (b) $y = \sqrt{x}$.

9. $y = x^3$.

10. $y = 4x^{1/3}$.

11. $y = \frac{32}{5}x^{3/5}$

Chapter IV.

1. If OR, OS are arms of an angle, take OB on OR, OC on OS, OB = OC = b. On OS, take OP = $a - b$; join PB; draw CX parallel to PB meeting OR in X; then OX = $b^2/(a - b)$.

2. Locus of C is a line XY parallel to UV, at distance 16 from it on the side opposite to RS.

3. If angle CUI = α_0, AV = m and $\frac{\sin\alpha_0}{\sin\beta_0} = \lambda_8$, then

$$\text{CI} = \lambda_8\lambda_7 m + \lambda_8 y - \lambda_8\lambda_7 x.$$

4. The locus is $x^2 + y^2 - 28x - 7y - 5 = 0$.

7. (c) same answer.

8. $x = 7$.

9. (b) $y = e^x - 1$.

10. (c) $y = -e^{-x} - \cos x - \sin x + x^2$.

Chapter V.

1. sub-normal = 3/2; sub-tangent = 8/3.

2. sub-normal = 10.

3. If P is the point, and S, H the two foci, the bisector of the angle between SP produced and SH is the tangent at P.

4. If G is (1/2, 0), line GP is normal.

5. sub-tangent = 2.

6. maximum is $x = 4$.

7. $a/e = 1/2$.

8. Derivative is $3x^2 + 4x$.

9. Derivative is $-1/x^2$.

Chapter VI.

1. **Hint:** $18 + 26i = (3 + i)^3$
   Answer: $x = 6$.

2. $31^2 + 12^2$; $4^2 + 33^2$; $24^2 + 23^2$; $9^2 + 32^2$.

7. There are so many ways of doing it; four of them are
   $22^2 + 50^2 + 23^2 + 6^2$; $30^2 + 32^2 + 33^2 + 20^2$;
   $52^2 + 5^2 + 12^2 + 26^2$; $26^2 + 39^2 + 26^2 + 26^2$.

Chapter VII.

10. **Hint:** First show that $2 < s_m < 3 - (1/2^{m-1})$.

11. **Hint:** (a) a in L gives $5-a^2 > 0$; take $b = a+(5-a^2)/(6+a)$. Show b is in L; so that L has no greatest.
(b) u in R gives $u^2 > 5$; take $c = \frac{u}{2} + \frac{5}{2u}$ and show that $c < u$ and $c^2 > 5$.

12. **Hint:** (a) a in L gives $p - a^2 > 0$; take $b = a + \frac{p-a^2}{(p+1)+a}$.
(b) Take u in R and show that $\frac{u}{2} + \frac{b}{2u}$ is in R.

Chapter VIII.

1. 3, 7.

2. The parts are 3, 7 or 6, 4.

3. $4 \cdot 8$.

4. son: 11/57 th part
first legatee: 11/57 th part; second legatee: 2/57 th part.

5. **Hint:** If each daughter's share is x, each son's share is $2x$;

first legatee's share is	$x - \frac{1}{5}(\frac{P}{3} - x)$
second legatee's share is	$x - \frac{1}{3}[\frac{P}{3} - \{x - \frac{1}{5}(\frac{P}{3} - x)\}]$
third legatee's share is	$\frac{P}{12}$

Answer: daughter's share = P/8; son's share = P/4; each legatee's share = P/12.

6. Put $x + y = t$ so that $z = 30 - t$
The second equation gives $x + 2t + 15 - (t/2) = 30$, so that $x = 15 - (3/2)t$
Equation (1) now gives $y = 30 - x - z$ so that $y = (5/2)t - 15$
z, x, y are positive: therefore $t < 30$, $t < 10$ and $t > 6$.
z, x, y are integers: therefore t is an even integer.
The only possible value of t is 8 giving $x = 3$, $y = 5$, $z = 22$.

7. 10 partridges, 15 doves, 75 sparrows.

8. **Hint:** Compute u_3, u_4, ... , u_9 successively.

9. **Hint:** Follow method of Section 10 used for finding x such that $x^2 + 5$ and $x^2 - 5$ are squares.
   Answer: $x = 25/4$.

10. $x = 13/2$.

11. **Hint:** Draw parabola $y^2 = 2(x-2)$ and hyperbola $xy = \sqrt{18}$.
    Answer: $x = 3$.

12. $x = 3$.

13. **Hint:** $-4 + 10\sqrt{-2} = (2 + \sqrt{-2})^3$.
    Answer: $x = 4$.

17. The correspondences

    (a) $\begin{matrix} 0 & 1 & 2 & 3 \\ 1 & 3 & 9 & 27 \end{matrix}$ and (b) $\begin{matrix} 0 & 1 & 2 & 3 \\ 1 & 27 & 9 & 3 \end{matrix}$

    are both distinct homomorphisms.

Index

Professor Dattatray B. Wagh

(1915 - 1994)

About the author

After his graduate degree from the Bombay University, Professor Wagh taught at the Ruia College, Bombay, for eight years, worked as Professor of Mathematics and Chairman of the Department at the Wilson College, Bombay, for fifteen years. He was selected to be the Principal of the first college in Goa and later as the Director of the Bombay University's Post Graduate Center of Instruction and Research at Panaji, Goa. He retired after a teaching career spanning 38 years.

He has written three text books on undergraduate mathematics; an experimental book on Calculus advocating that Integral Calculus be taught before its Differential counter-part, a text on Trigonometry and a text on Algebra. These books have run into several editions and in their own time, were the standard texts followed by all the colleges of Bombay University. In 1987, the birth Centenary of the great Indian Mathematician - Shrinivas Ramanujan, he wrote a book on the life and work of Ramanujan which was very well received by critics and public alike.

He was a patron of the Bombay Mathematical Colloquium and the guiding father of the Association. He was the first editor of the Bulletin of the Colloquium. He was also a Life-Member of the All India Mathematics Teachers Association.

His activities included many diverse facets of life. He was part of the first group of mountaineers from his state to scale the peaks of Himalayas. He was an avid fan of literature and actively organized literary conferences. He founded oraganizations that faught for teacher welfare, started new schools in rural areas and developed revenue sources for education of talented youth.

During his active as well as retired life, he was a constant guest speaker at various Universities, Mathematical meetings and Mathematical Conferences in India. Highly regarded teacher and mentor of a whole generation of mathematicians in Western India, a number of his students are now professors at various universities in India and United States.